Lecture Notes in Artificial Intelligence   12196

Subseries of Lecture Notes in Computer Science

More information about this series at http://www.springer.com/series/1244

Dylan D. Schmorrow · Cali M. Fidopiastis (Eds.)

Augmented Cognition

Theoretical and Technological Approaches

14th International Conference, AC 2020
Held as Part of the 22nd HCI International Conference, HCII 2020
Copenhagen, Denmark, July 19–24, 2020
Proceedings, Part I

 Springer

Editors
Dylan D. Schmorrow
Soar Technology Inc.
Orlando, FL, USA

Cali M. Fidopiastis
Design Interactive, Inc.
Orlando, FL, USA

ISSN 0302-9743   ISSN 1611-3349   (electronic)
Lecture Notes in Artificial Intelligence
ISBN 978-3-030-50352-9   ISBN 978-3-030-50353-6   (eBook)
https://doi.org/10.1007/978-3-030-50353-6

LNCS Sublibrary: SL7 – Artificial Intelligence

This Springer imprint is published by the registered company Springer Nature Switzerland AG
The registered company address is: Gewerbestrasse 11, 6330 Cham, Switzerland

Foreword

The 22nd International Conference on Human-Computer Interaction, HCI International 2020 (HCII 2020), was planned to be held at the AC Bella Sky Hotel and Bella Center, Copenhagen, Denmark, during July 19–24, 2020. Due to the COVID-19 coronavirus pandemic and the resolution of the Danish government not to allow events larger than 500 people to be hosted until September 1, 2020, HCII 2020 had to be held virtually. It incorporated the 21 thematic areas and affiliated conferences listed on the following page.

A total of 6,326 individuals from academia, research institutes, industry, and governmental agencies from 97 countries submitted contributions, and 1,439 papers and 238 posters were included in the conference proceedings. These contributions address the latest research and development efforts and highlight the human aspects of design and use of computing systems. The contributions thoroughly cover the entire field of human-computer interaction, addressing major advances in knowledge and effective use of computers in a variety of application areas. The volumes constituting the full set of the conference proceedings are listed in the following pages.

The HCI International (HCII) conference also offers the option of "late-breaking work" which applies both for papers and posters and the corresponding volume(s) of the proceedings will be published just after the conference. Full papers will be included in the "HCII 2020 - Late Breaking Papers" volume of the proceedings to be published in the Springer LNCS series, while poster extended abstracts will be included as short papers in the "HCII 2020 - Late Breaking Posters" volume to be published in the Springer CCIS series.

I would like to thank the program board chairs and the members of the program boards of all thematic areas and affiliated conferences for their contribution to the highest scientific quality and the overall success of the HCI International 2020 conference.

This conference would not have been possible without the continuous and unwavering support and advice of the founder, Conference General Chair Emeritus and Conference Scientific Advisor Prof. Gavriel Salvendy. For his outstanding efforts, I would like to express my appreciation to the communications chair and editor of HCI International News, Dr. Abbas Moallem.

July 2020   Constantine Stephanidis

.

HCI International 2020 Thematic Areas
and Affiliated Conferences

Thematic areas:

- HCI 2020: Human-Computer Interaction
- HIMI 2020: Human Interface and the Management of Information

Affiliated conferences:

- EPCE: 17th International Conference on Engineering Psychology and Cognitive Ergonomics
- UAHCI: 14th International Conference on Universal Access in Human-Computer Interaction
- VAMR: 12th International Conference on Virtual, Augmented and Mixed Reality
- CCD: 12th International Conference on Cross-Cultural Design
- SCSM: 12th International Conference on Social Computing and Social Media
- AC: 14th International Conference on Augmented Cognition
- DHM: 11th International Conference on Digital Human Modeling and Applications in Health, Safety, Ergonomics and Risk Management
- DUXU: 9th International Conference on Design, User Experience and Usability
- DAPI: 8th International Conference on Distributed, Ambient and Pervasive Interactions
- HCIBGO: 7th International Conference on HCI in Business, Government and Organizations
- LCT: 7th International Conference on Learning and Collaboration Technologies
- ITAP: 6th International Conference on Human Aspects of IT for the Aged Population
- HCI-CPT: Second International Conference on HCI for Cybersecurity, Privacy and Trust
- HCI-Games: Second International Conference on HCI in Games
- MobiTAS: Second International Conference on HCI in Mobility, Transport and Automotive Systems
- AIS: Second International Conference on Adaptive Instructional Systems
- C&C: 8th International Conference on Culture and Computing
- MOBILE: First International Conference on Design, Operation and Evaluation of Mobile Communications
- AI-HCI: First International Conference on Artificial Intelligence in HCI

Conference Proceedings Volumes Full List

1. LNCS 12181, Human-Computer Interaction: Design and User Experience (Part I), edited by Masaaki Kurosu
2. LNCS 12182, Human-Computer Interaction: Multimodal and Natural Interaction (Part II), edited by Masaaki Kurosu
3. LNCS 12183, Human-Computer Interaction: Human Values and Quality of Life (Part III), edited by Masaaki Kurosu
4. LNCS 12184, Human Interface and the Management of Information: Designing Information (Part I), edited by Sakae Yamamoto and Hirohiko Mori
5. LNCS 12185, Human Interface and the Management of Information: Interacting with Information (Part II), edited by Sakae Yamamoto and Hirohiko Mori
6. LNAI 12186, Engineering Psychology and Cognitive Ergonomics: Mental Workload, Human Physiology, and Human Energy (Part I), edited by Don Harris and Wen-Chin Li
7. LNAI 12187, Engineering Psychology and Cognitive Ergonomics: Cognition and Design (Part II), edited by Don Harris and Wen-Chin Li
8. LNCS 12188, Universal Access in Human-Computer Interaction: Design Approaches and Supporting Technologies (Part I), edited by Margherita Antona and Constantine Stephanidis
9. LNCS 12189, Universal Access in Human-Computer Interaction: Applications and Practice (Part II), edited by Margherita Antona and Constantine Stephanidis
10. LNCS 12190, Virtual, Augmented and Mixed Reality: Design and Interaction (Part I), edited by Jessie Y. C. Chen and Gino Fragomeni
11. LNCS 12191, Virtual, Augmented and Mixed Reality: Industrial and Everyday Life Applications (Part II), edited by Jessie Y. C. Chen and Gino Fragomeni
12. LNCS 12192, Cross-Cultural Design: User Experience of Products, Services, and Intelligent Environments (Part I), edited by P. L. Patrick Rau
13. LNCS 12193, Cross-Cultural Design: Applications in Health, Learning, Communication, and Creativity (Part II), edited by P. L. Patrick Rau
14. LNCS 12194, Social Computing and Social Media: Design, Ethics, User Behavior, and Social Network Analysis (Part I), edited by Gabriele Meiselwitz
15. LNCS 12195, Social Computing and Social Media: Participation, User Experience, Consumer Experience, and Applications of Social Computing (Part II), edited by Gabriele Meiselwitz
16. LNAI 12196, Augmented Cognition: Theoretical and Technological Approaches (Part I), edited by Dylan D. Schmorrow and Cali M. Fidopiastis
17. LNAI 12197, Augmented Cognition: Human Cognition and Behaviour (Part II), edited by Dylan D. Schmorrow and Cali M. Fidopiastis

http://2020.hci.international/proceedings

14th International Conference on Augmented Cognition (AC 2020)

Program Board Chairs: **Dylan D. Schmorrow, Soar Technology Inc., USA, and Cali M. Fidopiastis, Design Interactive, Inc., USA**

- Amy Bolton, USA
- Martha E. Crosby, USA
- Fausto De Carvalho, Portugal
- Daniel Dolgin, USA
- Sven Fuchs, Germany
- Rodolphe Gentili, USA
- Monte Hancock, USA
- Frank Hannigan, USA
- Kurtulus Izzetoglu, USA
- Ion Juvina, USA
- Chang S. Nam, USA
- Sarah Ostadabbas, USA
- Mannes Poel, The Netherlands
- Stefan Sütterlin, Norway
- Suraj Sood, USA
- Ayoung Suh, Hong Kong
- Georgios Triantafyllidis, Denmark
- Melissa Walwanis, USA

The full list with the Program Board Chairs and the members of the Program Boards of all thematic areas and affiliated conferences is available online at:

http://www.hci.international/board-members-2020.php

HCI International 2021

The 23rd International Conference on Human-Computer Interaction, HCI International 2021 (HCII 2021), will be held jointly with the affiliated conferences in Washington DC, USA, at the Washington Hilton Hotel, July 24–29, 2021. It will cover a broad spectrum of themes related to Human-Computer Interaction (HCI), including theoretical issues, methods, tools, processes, and case studies in HCI design, as well as novel interaction techniques, interfaces, and applications. The proceedings will be published by Springer. More information will be available on the conference website: http://2021.hci.international/.

General Chair
Prof. Constantine Stephanidis
University of Crete and ICS-FORTH
Heraklion, Crete, Greece
Email: general_chair@hcii2021.org

http://2021.hci.international/

Contents – Part I

AI and Augmented Cognition

Contents – Part II

Human Cognition and Behavior in Complex Tasks and Environments

Cognitive Modeling, Perception, Emotion and Interaction

User Evaluation of Affective Dynamic Difficulty Adjustment Based on Physiological Deep Learning

Guillaume Chanel[1](✉) and Phil Lopes[2](✉)

[1] Computer Science Department, University of Geneva, Geneva, Switzerland
Guillaume.Chanel@unige.ch
[2] Immersive Interaction Group, EPFL, Lausanne, Switzerland
phil.lopes@epfl.ch
http://cvml.unige.ch, http://iig.epfl.ch

Abstract. Challenging players is a fundamental element when designing games, but finding the perfect balance between a frustrating and boring experience can become a challenge in itself. This paper proposes the usage of deep data-driven affective models capable of detecting anxiety and boredom from raw human physiological signals to adapt the fluctuation of difficulty in a game of Tetris. The first phase of this work was to construct several emotion detection models for performance comparison, where the most accurate model achieved an average accuracy of 73.2%. A user evaluation was subsequently conducted on a total of 56 participants to validate the efficiency of the most accurate model obtained using two diverging difficulty adaptation types. One method adapts difficulty according to raw values directly outputted by the affective model (ABS), while another compares the current value with the previous one to influence difficulty (REA). Results show that the model using the ABS adaptation was capable of effectively adjusting the overall difficulty based on a player's physiological state during a game of Tetris.

Keywords: Affective computing · Machine learning · Physiology · Player modeling · Player experience · Autonomous difficulty adaptation

1 Introduction

The accessibility of games is often a widely debated topic, where some argue that games should be accessible for the majority of the player-base to enjoy, while others suggest that games should be challenging but fair. Balancing difficulty in digital games has always been a challenging task for designers, where specific game-play elements are adapted in order to achieve an ideal balance for a wide varying player skill-sets. Traditionally players can select the amount of difficulty

Co-financed by Innosuisse.

© Springer Nature Switzerland AG 2020
D. D. Schmorrow and C. M. Fidopiastis (Eds.): HCII 2020, LNAI 12196, pp. 3–23, 2020.
https://doi.org/10.1007/978-3-030-50353-6_1

they want to face during play (e.g. easy, medium or hard difficulty). However, this solution still presents a few limitations, as some players may not directly fit into any of the categories (e.g. high difficulty still too easy, or easy difficulty still too hard), or the game's actual difficulty progression during play may still be too steep leaving some players behind and frustrated, or even too slow making the game tedious.

A few games have attempted to address this problem by adjusting the difficulty of the game based on player performance. For example *Max Payne* (Remedy Entertainment, 2001) would slightly adjust the aim assist, provide additional ammunition and health kits when the player was having a hard or easy time to beat a particular level. The concept of a game reacting to a player's skill and progress is a fundamental idea of dynamic difficulty adjustment [4, 10, 32, 33], where the game attempts to manage mechanics, level layouts or the skill of non-playable opponents by using statistical game features to refine them towards the perfect challenge for the player.

Games are emotional experiences, they can induce a wide variety of emotions such as: frustration during challenging gameplay segments; anxiety in horror games; or even excitement. In fact, analyzing player reactions and emotions is of critical importance to evaluate games [27]. Thus, it makes sense for a game to also react to the players' emotions and not only to their performance. For this purpose, several researchers have proposed to develop models for automatic emotion recognition in games [7, 20, 36]. Unfortunately, the majority of research tends to be rooted specifically on the construction of accurate emotion recognition models, rarely addressing the question of game adaptation. In other words, a research question remain: how to employ emotion recognition models to adapt game difficulty and improve players' game experience?

Although this work mostly addresses emotion recognition for digital games, applications can be extended into other domains. Emotional models can allow systems to retain user engagement and attention by offering methods of detecting if a user is bored or frustrated, allowing the application to react accordingly and attempt to re-engage the user in some fashion.

This study attempts to answer this question by extending the work presented in [7]. It is based on models trained from physiological signals and which are capable of detecting anxiety and boredom during play. Given that most notable works using physiology to model affect in games have used skin conductance with a wide degree of success [21, 25, 36], this study applies deep learning to this modality for emotion recognition. The emotional information is then used to adapt the difficulty of a Tetris game. Two types of adaptation methods are explored, where one type directly uses the model decision, while the other compares the previous model output with the current one and then adjusts the difficulty accordingly. A user study was conducted exploring the viability of both approaches during a real-time gameplaying session.

2   Related Work

The following section provides literature of previous work exploring dynamic difficulty adjustment and affective modeling within the domain of digital games.

2.1   Dynamic Difficulty Adjustment

The concept of challenge and difficulty within the domain of digital games has always been a heavily debated topic, where several theories have been formulated. In particular the theory of flow [9], has been widely applied as both a design paradigm [16] and for Dynamic Difficulty Adjustment (DDA) systems [7]. DDA systems are capable of adjusting and eventually personalizing the difficulty during play. Thus, the ideal DDA system would still provide players with a degree of challenge, such that it is not overwhelmingly hard for the player but still challenging enough for the player to feel accomplished and engaged in the experience. According to the flow theory, a too high (respectively low) challenge should lead the player towards an emotional state of anxiety (respectively boredom). In [7] the concept of flow is used to balance difficulty by measuring players' anxiety and boredom: if the player is anxious (resp. bored) then reduce (resp. increase) difficulty.

One of the most common approaches towards designing DDA systems is through the measurement of performance metrics (e.g. number of deaths, damage received, damage done). One of the first academic example of DDA optimized adversarial agent strategy based on a function of challenge, by using an online evolutionary method [10]. Adversarial agents adapt their "ability" based on player's performance, which is calculated through specific heuristics derived from statistical game parameters (e.g. rate of missed or hit shots, games won and lost, time to finish a task). More recently, reinforcement Learning and evolutionary computation strategies for DDA were compared on a simplified racing game. A more data-driven approach was applied in [32], where the authors attempted to model player experience paradigms such as fun, frustration and challenge for the generation and personalization of *Super Mario Bros.* (Nintendo, 1985) levels.

As noted in [27], affective experiences are a fundamental part of digital games and should be one of the main evaluation factors during play-testing and game evaluation phases. Thus, it makes sense to approach the problem of difficulty management through the perspective of emotion, and player experience. This paper takes an affective computing approach exclusively, rather than using in-game performance metrics, by using physiological monitoring (i.e. Galvanic Skin Conductance) for the measurement of player affect.

2.2   Affective Modeling in Games

In the field of affective computing, games are often used as stimuli to induce players' emotional reactions which are then recorded and used to create emotion recognition models [2,21]. The detection of an individual's emotions can be achieved through several means, such as facial expressions and vocal prosody [6], but also by using physiological signals [15, 26]. In order to target applications such as video games, a noticeable shift has been observed in the last decade towards assessing emotions in more natural and realistic situations [3, 30]. This shift has demonstrated that the majority of algorithms suffer from a drop of performance when compared to the recognition of controlled and acted emotions. In addition, emotion recognition models can be trained to recognize any player's emotions (i.e user independent) or only those of the player for which the models were trained (i.e. user dependant) [26]. It is important to note that user independent models have the advantage that they do not necessitate any learning phase prior to play but they generally suffer from a significant drop in performance [3].

In the context of emotion recognition from physiological signals for video gaming, Rani et al. [29] proposed to classify three levels of intensity for different emotions, including anxiety, using several physiological signals measuring heart activity, muscle activity, electro-dermal activity and skin temperature. The best average accuracy obtained with this method was 86%. Physiological user-independent models for affect recognition in games were proposed in [7]. By combining peripheral physiological signals (electro-dermal activity, heart rate, skin temperature) with brain signals a performance of 63% was obtained to distinguish three emotions: boredom, engagement and anxiety. The current paper is a significant improvement of this work, which in particular now uses deep networks to improve performance and evaluates players perception of affective difficulty adaptation. A deep convolutional network was used in [22] to detect flow, boredom, and stress from heart rate and electrodermal activity during Tetris play (15.5 h of data). The results demonstrate that user-independent models can achieve an accuracy around 70% when classifying two classes (stress vs. not stressed and flow vs. not flow). However the boredom state is the most difficult to detect with an accuracy of 57% when classifying boredom vs. other states. In addition, when flow was detected, the players obtained a better score than in the other emotional states showing that physiological flow can be an indicator of performance. Interestingly, a remote approach to physiology measurement have also been adopted to detect stress and boredom in games with an accuracy of 61.6% [5]. Remote physiological sensing consist of measuring physiological signals such as heart rate without using contact sensors [35] (e.g. a webcam is used to measure the slight changes of skin color produced by heart rate variability). The results also showed that despite of a small drop of performance compared to contact sensors, fusing physiological remote sensing with facial modalities was increasing emotion recognition performance compared to any model using a single modality.

Affective computing can bring substantial benefits to the domain of digital games, allowing designers and the game itself to construct personalized experiences for players based on predictions of their current emotional state. For example in the work of Lopes et al. [20], affective models were used to place sounds within a procedurally generated level to follow a specific emotional fluctuation for the player to experience. To our knowledge, emotion classifiers using physiological signals as inputs have been evaluated for DDA in two studies [11,19].

The classification methods employed in [19] were the same as those presented in [29] but classifiers were retrained for the new players. The reported accuracy dropped to 78% but a user-study showed a significant improvement of player experience compared to difficulty adjustment based on performance. Although this demonstrates the interest of using affective computing for the purpose of game adaption, the proposed model was user-dependent and required a one hour training session for each new player in order to tune the classifiers. In [11], the difficulty of a Tetris game was adapted depending on players' brain activity and using three adaptive strategies: conservative, moderate and liberal (ranging from the one presenting less adjustments to the one providing a lot of adaptations). In this experiment, players reported a higher alertness when playing with the conservative version compared to the liberal one, showing that adaptive systems should be designed with care and not overly adapt to the players physiology.

Even though a substantial amount of research exists for the construction of affective models, works that take these models and effectively integrates them within a game for user evaluation remain scarce. This study argues that although the conceptualization of affective models is an important process, it is also important to think and conceptualize what to do with the information obtained from such models and how to directly apply them within the applicative context. This work investigates methods for directly applying an emotion recognition model in games by observing both: the players actual perception of difficulty during play, and the actual behavior of said model during several play tests. Furthermore, this paper also explores two different adaptation methods which change difficulty according to either: the raw values obtained from the model (ABS); or by comparing the previous and current output to make a decision (REA).

3   Methodology

This section details the experimental methodology utilized for the study of adaptive difficulty adjustment using affective modeling. In a first step, a model able to recognize player's emotions was developed. It followed the physiological data driven approach proposed in [7,12]. However, instead of extracting features from the recorded signals we hereby propose an end-to-end learning solution based on the combination of deep convolutional [24] and Long Short Term Memory (LSTM) networks [13]. The proposed method was compared to the original method in [7] by using the same dataset. In a second step, the trained system was employed to test two difficulty adaptation strategies in real-time. These two strategies were tested and compared in a user study where the players experience was collected through self-reports.

3.1   Affective Data

The data collection process used for this study is fully described in [7] and summarized below. The affective dataset was acquired by measuring the physiological activity of 20 participants playing a Tetris game for 6 sessions of 5 min each (10 h of data). Each game session was played at a given difficulty level in order to elicit the following emotions: boredom (difficulty level lower than their skill), flow (difficulty level that matches their skill) and anxiety (difficulty level higher than their skill). The players' skill was determined on previous play sessions. Each condition was played twice hence the 6 sessions. For this work only the boredom and anxiety emotional states are considered because: (i) the flow class was the most difficult to classify as reported in [7], (ii) by performing adaptation based on boredom and anxiety we expect that the difficulty level will oscillate around the ideal player difficulty. In addition, while [7] addressed the importance of fusing several physiological signals to improve performance, we preferred to adopt a user friendly interface which relies only on the most efficient physiological signal: electrodermal activity.

3.2   Electrodermal Activity

The Electrodermal Activity (EDA) is a measure of the sympathetic nervous system, specifically the sweat gland activity. An increase of activity in sweat glands usually occurs when one is experiencing arousing emotions such as stress or surprise [17,31]. In the original dataset, electrodermal activity was recorded using the Biosemi Active 2 system with a sampling rate of 256 Hz which was later undersampled to 8 Hz (EDA is a slow varying signal). The Biosemi Active 2 electrodes are difficult to equip and we therefore favored an open source solution for the user study. We used the sensors proposed in [1] and inspired from [28], which measured EDA (more precisely skin resistance) at a sampling rate of 8 Hz. For both experiments the same positioning of electrodes was used: two electrodes positioned on the medial phalanges of the index and middle finger.

Signals of each session were segmented into 20 s windows with an overlap of 19 s. This specific window duration was chosen as a compromise between a fast adaptation time and the necessary duration required to reliably infer user emotions. This overlap also provides a high number of samples which are generally necessary for the training of deep networks. This windowing procedure led to a total of 21918 windows with a slight skew towards anxiety (48.7% of windows belonging to the anxiety class), as it was necessary to discard two sessions due to recording errors.

3.3   Baseline Affective Models

In order to create baseline models (i.e. models to which the deep approach will be compared), the most relevant EDA features suggested in [7] were selected to characterize both the Phasic and Tonic components of an EDA signal. The average EDA was computed to reflect the Tonic component. The percentage of samples which included an increase of the EDA signal was used to indicate the importance and duration of conductance peaks. Lastly, the total number of peaks within an EDA signal was used.

All features are computed for each time window as described in Sect. 3.4, and concatenated into a feature vector. Due to an abundance of inter-participant and inter-session variability when monitoring physiological signals [37], it is standard practice to record participants during a resting period, and subsequently subtract these measurements from the signal obtained during the actual activity. For this study, a resting period of 20 s was recorded before each play session. A rest feature vector was constructed based on the rest signals and subtracted from all the play feature vectors of the same session.

Results from the previous study [7] suggest that discriminant analysis classifiers are superior on the proposed dataset. Thus, this work compares the proposed end-to-end learning approach to that of a linear discriminant analysis (LDA) and a quadratic discriminant analysis (QDA). Although the linear approach might show better generalization on the test set, the quadratic method allows to define non-linear boundaries between classes as is the case of the proposed deep neural network.

3.4   Affective End-to-End Learning

The advantage of using deep neural networks is that it offers the possibility to train models capable of identifying the most useful features of a signal and build a representation of the problem through intermediary layers [18]. In other words, instead of including expert knowledge to define and select relevant features used for classification, the raw signals themselves are fed directly to the network, building a discriminative internal representation of the data. For this work, the derivative was calculated for each window of a signal and then used as input for the networks, so as to reduce inter-participant variability which tends to occur due to varying data range (i.e. the general amount of sweat on the fingers). Contrarily to the baseline methods presented in Sect. 3.3 it does not need to rely on the recording of a rest period.

The following deep architecture, inspired from [34] and stacking convolutional layers with a LSTM layer, is proposed to model EDA:

- 1D convolutional layer with 16 kernels of size 2, a stride of 1, no padding and a linear activation function;
- max-pooling layer with size 2;
- the succession of 3 1D convolutional layers (each with 8 kernels of size 2, a stride of 1, no padding and a linear activation function) and max-pooling layers of size 2;
- LSTM layer with 16 recurrent neurons and a softsign activation function;
- 1 neuron fully connected layer acting as a logistic regression (i.e. a sigmoid activation function with a loss measured using binary cross-entropy).

With this neural architecture the receptive field of a neuron in the last convolutional layer corresponds to 31 samples of the input signal. This implies that a sample outputted by a neuron of the last convolutional layer corresponds to approximately 3.9 s of the input signal. This duration and the corresponding architecture was specifically chosen as it is enough to detect an EDA peak. In other words, the convolutional layers are expected to act as peak detectors and characterizers. The LSTM takes these temporal features as inputs and memorizes their temporal occurrence. The LSTM layer can be seen as a peak counter with the capabilities of establishing a temporal relationship among them.

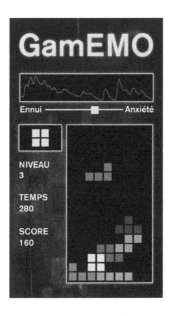

Fig. 1. Screenshot of the affective Tetris game. The upper window shows an example of an EDA signal with a duration of 70 s, which was recorded during the user study. Several galvanic responses (peaks) can be observed in the signal.

The network was implemented using Keras [8]. The decay rate was set to 10^{-4}, while the momentum to 0.5. The models were trained with 100 training epochs and a batch size of 2000. The performance of every classifier was evaluated using a leave-one participant out cross-validation (see Sect. 4.1). In addition, 70% of the training data was used to adjust the network weights while 30% of the data was used to control the validation error of each fold. Since all validation curves showed a plateau (even with a number of epoch higher than 100) it seemed that the model did not overfit the training data. The last trained model at epoch 100 was thus chosen as the model to be tested. Finally, in order to improve the performance of this model, we also performed ensemble learning by combining 20 neural networks with the same architecture. These 20 networks were trained on splits of the training set containing each 70% of the data randomly selected from the full training set. The decisions of these models were then averaged to take a final decision.

3.5 The Tetris Game

In Tetris Fig. 1, the difficulty is mainly controlled by the falling speed of the Tetrimino pieces. The simplicity of this particular difficulty mechanic is the main reason why Tetris was chosen, as it allows our adaptive models to easily manipulate the speed by which a chosen piece is falling. In addition, the Tetris game can be played with only one hand which is convenient to place the EDA sensor on the non-dominant hand.

The main difference between traditional Tetris and the version utilized within this work, is that the game will not terminate once a block reaches the upper "game-over" boundary, and instead will clear all the blocks and instantly reset the playing field. This was done to have play sessions of the same duration. Furthermore, the game also logs player and model information that occurs during play. All events are logged with an adjoining time-stamp and include: the predicted emotion (boredom vs. anxiety); horizontal line completion, and the moment of a game-reset (i.e. upper boundary reached).

3.6 User Study

To validate the effectiveness of our DDA approach a user-study was conducted with the Tetris game. The objective of this investigation was to observe the behavior of the emotion recognition model during an entire game of Tetris and how the difficulty was perceived by each participant during play. The experiment was conducted in a public setting during an open science and research demonstration day. Thus, it was tested in a ecological situation where emotion recognition models generally performs worst than in a laboratory (for instance players are playing in front of many people which can bias physiological reactions due to the emerge of social emotions). Two computers were used, where each one was running a different version of the Tetris game: the ABS and the REA adaptation variants, which are further described below. Playtime was limited to 120 s. All participants gave their informed consent, while the EDA sensors were being prepared. At the start of each game both the REA and ABS versions would equally cycle through 3 diverging starting difficulties: *Easy*, *Medium* and *Hard*. This was done in order to visualize the efficiency of each adaptation type towards situations of extreme difficulty, or simplicity. During play, all participants were required to use noise-canceling headphones to avoid physiological reactions due to the environment. At the end of the play session each participant was asked to rate on a five point Likert scale the perceived difficulty of the first and last 60 s of the game.

The proposed affective models output a value between the interval $[0, 1]$, where values close to 0 indicate a strong certainty on boredom while values close to 1 represents a higher confidence towards anxiety. Classifiers outputs were computed using 20 s of signal. The first output was thus obtained 20 s after the play begun. However, using a signal buffer, the following decisions were taken every 10 s based on the last 20 s of signal. Based on classifiers outputs two game adaptation strategies were developed. The first strategy called absolute (ABS) consists of increasing the game difficulty if the classifier output is lower than 0.5 (i.e. boredom is detected), or decreasing it if otherwise (i.e. anxiety is detected). By using this method it is expected that difficulty will start to converge towards an "ideal" challenge for players, lying within the boundary between boredom and anxiety. The second strategy referred to as relative (REA), consists of changing the difficulty at step t with respect to the previous classifier output at step $t - 1$. If the classifier output is superior to the previous one, than adaptation assumes that the player is more anxious and thus decreases the difficulty, while if the contrary occurs the adaptation assumes the player is bored and thus increases difficulty. The idea behind this strategy is that the model would utilize historical information to make the decision to either increase or decrease the overall difficulty by simply comparing output values rather than use the absolute model value. Thus, with this adaptation style we consider that the fluctuation of boredom/anxiety directly relates to the previous measurement and not a global scale. Given the short length of a playing session, using the direct previous value $(t - 1)$ was considered to be sufficient.

4   Results

4.1   Affective Modeling

Models previously presented in Sect. 3 are trained and subsequently tested on the dataset described in Sect. 3.1. Each classifier is cross-validated using a leave one participant out method. The classifiers performance is presented in Table 1. Three performance measures are used: the accuracy measures the percentage of samples correctly classified, while Cohen's Kappa and F1 scores are measures which are not sensitive to the slight class imbalance present in the dataset.

Table 1. Classifiers performance for anxiety and boredom recognition

Classifier	Accuracy	Kappa	F1
QDA	66.1%	0.315	0.647
LDA	69.5%	0.388	0.693
DeepNet	70.4%	0.408	0.704
DeepNet ensemble	73.2%	0.465	0.732

Table 1 showcases the performance of each classifier trained on the data collected in a laboratory. Results suggest that the proposed deep network performed just as efficiently, and even slightly better than the baseline models, achieving an accuracy of 69.5% and 70.4% for LDA and the deep network, respectively. In addition to validating the superiority of the Deep network model, the kappa and F1 score demonstrate that none of the classifiers were sensitive to the slight class imbalance.

Given the favorable results obtained from deep networks, an ensemble of these networks was also constructed to better generalize on the test data. Overall, an improvement of 2.8% was observed when using ensemble learning on the deep networks, showcased in Table 1. Table 2 shows the confusion matrix of this model. A slight bias can be observed towards the anxiety class, which suggests that this network tend to over-estimate anxiety. Taking into consideration the improved performance of this model, it was decided to integrate it specifically in the subsequent user studies for the detection of boredom and anxiety during a playthrough of Tetris. For this use case, the ensemble deep network architecture was re-trained on the full data set (i.e. without cross validation)

Table 2. Confusion matrix (in %) of the ensemble of deep networks

	Estimated boredom	Estimated anxiety
True boredom	36%	15%
True anxiety	12%	37%

4.2 User Study

Although over 100 individuals participated in the experiment, it was necessary to discard almost half of this collected data because signals were quite noisy and several participants were below the age of consent (i.e. minors). In total 56 of the participants presented usable data, where 26 and 30 of these were assigned to both the ABS and REA variations of the game, respectively. Table 3 showcases the distribution of the starting difficulty according to each adaptation method experimented.

Table 3. The total number of participants and their respective Tetris variation and starting difficulty.

Difficulty	ABS	REA
Easy	13	14
Medium	5	9
Hard	8	7
Total	26	30

In terms of familiarity with the Tetris game, the majority of participants stated that they had frequently played the game (32%), while 14% and 20% claimed to have never or just occasionally played the game. 14% asserted that they were playing regularly, while the remaining participants did not directly state their skill levels.

Decision Accuracy During Play: A Binomial significance test was conducted to check if the difficulty of the second half was harder than the first half. The most difficult half of the game was determined by selecting the half with the highest self-reported difficulty on the likert scales (see Sect. 3.6). This ranking approach was used to remove participant bias, as the interpretation of such scales may vary substantially among different individuals [14]. Furthermore, participants who gave the same rating for both halves of the game were discarded from the analysis. The significance of the result was tested using a binomial test.

Figure 2 shows that there is a discrepancy between the ABS and REA adaptation methods. For REA the majority of participants (80%, $p < 0.05$) ranked the second half of the game as more difficult than the first half. Early impressions from these results suggest that the REA adaptation method was not particularly effective in managing the overall difficulty of the participants, with the majority of individuals struggling with the second half of the game. Contrarily to the REA results, participants presented more ambiguous answers for the ABS version of the game, with a slight non-significant bias towards the second half being considered more difficult. Unlike REA, the ABS results show that participants had a more balanced experience with this version of the game, although these specific results were not be significant it does suggest that this particular model adapted more accurately towards the players skill.

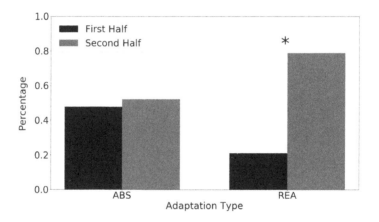

Fig. 2. Distribution of difficulty (i.e. first and second halves) chosen by participants, according to the different adaptation variants. Significant results are indicated by *, where $p < 0.05$.

Table 4. The classification accuracy of the model in the user-study, computed for participant self-reports, using different adaptation methods for each starting difficulty game variant (i.e. Easy, Medium and Hard). p-value is indicated in parenthesis

Adaptation	Acc_{Easy}	Acc_{Medium}	Acc_{Hard}
ABS	**31.6% (0.008)**	69.6% (0.09)	62.9% (0.18)
REA	51.7% (0.89)	64% (0.23)	55% (0.83)

To measure the detection performance in the user-study according to participants reports, we defined that the classifier performed well if it increased (resp. decreased) difficulty in the second half when the participants reported a low (resp. high) difficulty in the first. To facilitate the analysis and concentrate specifically on how the classifier adjusted difficulty based on an initial perception, runs where participants perceived the first half of the game to be "okay" were discarded. Table 4 shows the accuracy of the model for each starting difficulty game variant. The ABS adaptation type presents the most varied classification accuracy, where the easy difficulty variant obtains a statistically significant average accuracy of 31.6% ($p < 0.05$), a substantially lower accuracy when compared to both medium (69.6%, $p < 0.1$) and hard (62.9%, $p < 0.2$) variants. No significant results were obtained for the REA adaptation.

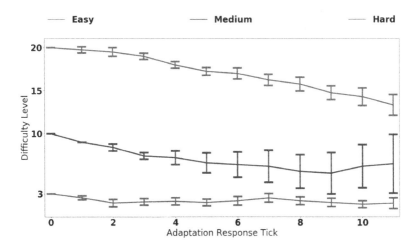

Fig. 3. Difficulty fluctuation during play using the ABS model. Each tick represents the average and standard error of each increase or decrease in difficulty level done by the model. The different colors represents the game starting difficulties. (Color figure online)

Difficulty Adaptation over Time: During play both models attempt to slowly adapt the player difficulty based on the physiological responses obtained from the player. In order to visualize how the models adapted play, the average difficulty of the game (and the standard error) was plotted in Fig. 3 and Fig. 4 for each type of adaptation and starting difficulty. In these figure, each tick of the x-axis represents a change of difficulty done according to the detected player's emotion. The difficulty (y-axis) is represented as a positive natural number, where a higher level represents a more difficult state of the game. Unfortunately, due to an error with the logging system within the REA version, we were unable to record the final level modification of this version of the game.

Figure 3 shows that the ABS adaptation did influence the difficulty fluctuation during play. For the *hard* condition in particular the ABS model detected anxiety and progressively lowered the difficulty until almost reaching the "medium" starting difficulty. Interestingly, in the *medium* condition a similar pattern is observed to that of the *hard* condition, where the early decisions of the model is to consistently lower the difficulty level. However, contrarily to the previous difficulty during the mid-game a larger standard error is observed, and during the latter stages the average difficulty actually increased. This does suggest that for some participants the adaptation did not just constantly reduce the difficulty, in fact during the latter stages of the game the model attempted to adjust the overall game difficulty, which makes sense considering that players potentially proved more proficient in the latter portions of the game. Out of all game variants the *easy* condition proved to be the most stable, with little fluctuation throughout the game. This goes in-line with the low accuracy results observed in this condition and further suggests the models struggle in increasing difficulty (i.e. detecting the boredom state).

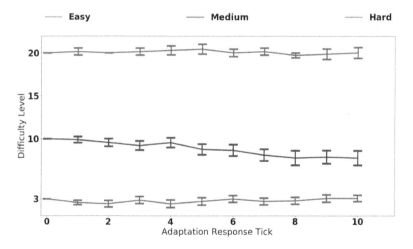

Fig. 4. Difficulty fluctuation during play using the REA model. Each tick represents the average and standard error of each increase or decrease in difficulty level done by the model. The different colors represents the game starting difficulties. (Color figure online)

Contrarily to the previous adaptation type, for the REA plot (Fig. 4) a more stable difficulty fluctuation can be observed across all game variants. The *medium* condition presented the most fluctuation, with the game slightly reducing the overall game difficulty during the second half of play, which may be a reason why a higher accuracy was previously observed for this condition compared to the other two. Both *hard* and *easy* conditions did not present substantial differences, where for the majority of play the adaptation method does not change the difficulty of the game. The reason for this stabilization is due to how this model operates, specifically when comparing the previous and current outputted values by the model for the decision making process. More precisely, this adaptation method attempts to treat outputted values as a relative measure according to the previous one, even if the difference between the two values are minimal a decision is still made. This consequently manifests as a constant state of difficulty fluctuation that does not allow the adaptation to commit towards a "long-term" objective, and thus why it often gets stuck within a range of around three levels from the initial starting difficulty.

5   Discussion

5.1   Model Performance

The performance obtained from the classifiers trained on laboratory data show the superiority of the deep network architecture. As previously mentioned in Sect. 3.3, there are two factors which increase the performance of LDA/QDA: feature selection and the rest period subtraction. Feature selection was done according to our previous study, which was accomplished on the same dataset [7]. Thus, it is possible that the baseline models slightly over-fitted the database using knowledge extracted from the full dataset. This could result in artificial improvement of the QDA/LDA performance. In addition, the LDA/QDA used additional information (i.e. the rest period) to take decisions on anxiety and boredom. However, despite these potential advantages of the baseline models, the deep network still reached an higher accuracy than both QDA and LDA. Importantly, avoiding a rest period allows deploying emotion recognition systems without any calibration phase to the user. This is particularly relevant to reach market deployment or to perform experiment in an ecological situation such as the one we presented.

It is important to note that the data used for training was collected from a total of 20 participants, aggregating only 590 min of total play-time. Although the number of samples is relatively high (21918 samples), there is a lot of data redundancy since consecutive windows overlap by 95%. This contrasts with the typical tendency observed in previous work on deep networks, where such models are often trained on large subsets of data, which typically is a requirement for these types of network architectures [23]. This demonstrates the capabilities of the proposed network in generalizing on unseen data without requiring of huge quantities of training data.

It remains difficult to compare the obtained results with other studies because the number and type of emotions to be detected is not the same. However, the Kappa score of 0.465 obtained by the ensemble of deep networks demonstrates the feasibility of user-independent affect recognition contrarily to the 0.01 Kappa score obtained in [3]. Although the best reported accuracy of 73% (for two classes) is lower than the 78% (for three classes) reported in [19], our model is user-independent contrarily to the one proposed in [19]. The closest study to our work is probably [22] as anxiety, boredom and flow were detected using a deep architecture trained on 15.5 h of physiological signals (heart rate and EDA). A very similar performance was obtained using this approach (around 70% to classify 2 classes). Together with our results, this reinforces the statement that deep architectures can be trained with only a 10 to 15.5 h of physiological activity.

Finally, the performance reported by the players in the user study is much lower than the one computed on the original database, dropping to a value between 31% to 70%, depending on the original difficulty. The low performance in the easy difficulty could be explained by a bias of the classifier toward the anxiety class which is already observed in the confusion matrix computed on the laboratory data (see Table 2). Interestingly a similar bias was reported in [22] who showed that boredom is more difficult to detect than flow and stress. However this bias might not be the only responsible for the drop of performance. An additional explanation is that the EDA sensors were not the same for training and for the real time measurements. It is thus possible that the features captured by the deep networks on the original sensors do not have the same shape and occurrence in the signals recorded by the real-time sensors. To circumvent this problem, a transfer learning approach [37] might allow to adjust the results to the new sensors.

5.2   User Perception of Adaptation Methods

Early results show a clear distinction between both the ABS and REA models. The ABS adaptation in particular also showed a few distinctions between the different game difficulty conditions. For the *hard* game type, the ABS adaptation method did attempt to adjust the difficulty towards more sensible levels considering that the majority of participants were mostly casual players, which can be observed from the degree of accuracy obtained (62.9%). Although the latter results were not significant, it important to note that for both *hard* and *medium* conditions a lower amount of participant data was collected. The tendency for ABS was to guide participants towards the *medium* level difficulty interval, as the model attempted to gradually lower the difficulty as much as possible (i.e. one level per "tick"). Furthermore, for this particular difficulty variant the ABS adaptation presented a clear bias towards decreasing the game level (i.e. detecting anxiety), which may be the reason why this model performed better in the *Hard* compared to *Easy* condition. In a way this does show that the model is capable of self-regulating the difficulty based on the participant to a certain degree, however these initial results suggest that for the model to do so efficiently the participant must already be sufficiently challenged.

For example the *Medium* version of the game presents an increased variability in the standard error, compared to the other game difficulty variants, in the latter portions of the game. This can be expected when managing difficulty within this range, as the majority of players may feel more comfortable playing the game. The range itself is also complicated to measure effectively, as it lies exactly within the space of being "just right" and, either "too difficult" or "too easy". Interestingly, this game condition presented a similar pattern to *hard* during the initial instants of the game, but in latter portions ABS did attempt to re-increase the difficulty. This could be explained by the fact that the game effectively adjusted to the player's competence and the plateau observed at the end of the curve corresponds to the average competence of the recorded players. For the *easy* condition on the other hand it was apparent that the ABS adaptation model struggled in comparison due to the present anxiety bias. There are several instances where the model attempts to lower the difficulty even though the game was not considered challenging at that point in time. This suggests that the model does have a higher difficulty in detecting the boredom state in relation to anxiety as explained in Sect. 5.1.

Unlike the ABS adaptation, the REA variation presented a more stable difficulty fluctuation in comparison. This phenomenon was observed on all three starting difficulty variations, which suggests that on average the model lingered within the same difficulty range over the course of a playing session. This is also reflected on the average accuracy of the REA adaptation method, which in general presented ambiguous accuracies ($\approx$55%). This lack of adaptation was also apparent to the participants themselves, where a substantial bias on perceived difficulty is present specifically on the second half of a playing session. The reasoning for this bias is due to the game itself and how it tends to be played. At the latter points of the game players tend to have a more Tetrimino cluttered playing area due to previous player mistakes, this will consequently slowly increase the chances of a potential game-over situation, which may be perceived as the game becoming substantially more difficult.

5.3   Future Work

Given that the current adaptation methods have only two options, i.e. either increase or decrease difficulty, a potential third class might be advantageous for ambiguous values outputted by the model. For future approaches on adaptation, including an hysteresis on the change of decision might actually help mitigate some of the bias that may exist within the model towards anxiety. Furthermore, for the ABS adaptation it might make sense to discard predictions that are within close proximity of 0.5 (i.e. the boundary between both emotional states), as they are too ambiguous to make an actual decision on difficulty.

Although, this paper focused specifically on experimenting with two diverging types of difficulty adjustment systems, an alternate study focusing specifically on the enjoyability while using such a system might be advantageous, where a comparison can be made between the classical approach (i.e. classic Tetris) and adaptive difficulty. A common argument against these types of dynamic adjustment systems usually focus on how certain players actually enjoy overcoming these arduous challenges; offering the player a sense of accomplishment. Although this is certainly true for certain types of games, this does not discount the advantages that such adaptive systems may offer in other applications for which having a task challenge that matches users' competences is relevant.

Future improvements to the proposed models could also possibly be obtained if in addition to physiological data, information from the gameplay itself could be utilized as input. In the current version of the model, input data consist purely of raw GSR data. The accuracy of these predictions could potentially be enhanced if in conjunction to the physiological data, some form of gameplay information could also be processed into features informing the model of: current difficulty level, number of lines, Tetriminos used and among others. However, in order to pin-point exactly what features influence emotion, and how to specifically train a model to use these gameplay features and relate it to affect is not a straight-forward task. The simplest method would be to simply have participants play a game of Tetris, while logging all gameplay elements and measuring GSR synchronously. From this data, attempt to correlate certain gameplay characteristics to specific traits obtained from GSR signals (i.e. phasic peaks). Another method could be through real-time annotation, where participants annotate replays of Tetris playing sessions through an annotation system like RankTrace [21]. From these annotations a relation could possibly be made from the gameplay features and the annotated data.

Lastly, it is important to keep in mind that the models explored within this work can also be applied to several other domains. For example, applications that require constant engagement and monitoring from their users can benefit by reacting accordingly if an individual's attention starts to wane due to boredom. The software may be able to thus stimulate such individuals in order to retain their attention to the task at hand.

6   Conclusions

This paper presented a comparison between two diverging approaches for an affective model to adapt difficulty within the Tetris game. The affective model is capable of adjusting difficulty using human physiological responses, specifically electrodermal activity, and making an emotional prediction of the player's state, i.e. anxious or bored. Several algorithms were tested, where a higher performance was observed with an ensemble deep neural-network approach (73.2%). This model was then tested with 56 participants using two different adaptation approaches: the absolute (ABS) and relative (REA). Results show that the ABS outperformed the REA method. The ABS model was particularly efficient at adapting the game difficulty when players were stating to play in hard or medium difficulties. This paper builds upon previous research in dynamic difficulty adjustment by testing the affective models out of the laboratory.

Acknowledgments. We would like to thank the Blue Yeti (http://www.blueyeti.fr/en/) company for developing a first version of the affective Tetris game which was adjusted for the purpose of this study. Co-financed by Innosuisse.

References

1. Abegg, C.: Analyse du confort de conduite dans les transports publics. Master thesis, University of Geneva (2013)
2. Alhargan, A., Cooke, N., Binjammaz, T.: Affect recognition in an interactive gaming environment using eye tracking. In: 2017 7th International Conference on Affective Computing and Intelligent Interaction (ACII), pp. 285–291. IEEE (2017)

3. Alzoubi, O., D'Mello, S.K., Calvo, R.A.: Detecting naturalistic expressions of non-basic affect using physiological signals. IEEE Trans. Affect. Comput. **3**(3), 298–310 (2012). https://doi.org/10.1109/T-AFFC.2012.4
4. Andrade, G., Ramalho, G., Santana, H., Corruble, V.: Extending reinforcement learning to provide dynamic game balancing. In: Proceedings of the Workshop on Reasoning, Representation, and Learning in Computer Games, 19th International Joint Conference on Artificial Intelligence (IJCAI), pp. 7–12 (2005)
5. Bevilacqua, F., Engström, H., Backlund, P.: Game-calibrated and user-tailored remote detection of stress and boredom in games. Sens. (Switz.) **19**(13), 2877 (2019). https://doi.org/10.3390/s19132877
6. Calvo, R., et al.: Introduction to affective computing. In: The Oxford Handbook of Affective Computing, pp. 1–10. Oxford University Press (2015).https://doi.org/10.1093/oxfordhb/9780199942237.013.040
7. Chanel, G., Rebetez, C., Bétrancourt, M., Pun, T.: Emotion assessment from physiological signals for adaptation of game difficulty. IEEE Trans. Syst. Man Cybern. Part A Syst. Hum. **41**(6), 1052–1063 (2011). https://doi.org/10.1109/TSMCA.2011.2116000
8. Chollet, F., et al.: Keras (2015). https://github.com/keras-team/keras
9. Csikszentmihalyi, M.: Beyond Boredom and Anxiety. Jossey-Bass, San Francisco (2000)
10. Demasi, P., Adriano, J.D.O.: On-line coevolution for action games. Int. J. Intell. Games Simul. **2**(2), 80–88 (2003)
11. Ewing, K.C., Fairclough, S.H., Gilleade, K.: Evaluation of an adaptive game that uses EEG measures validated during the design process as inputs to a biocybernetic loop. Front. Hum. Neurosci. **10**, 223 (2016). https://doi.org/10.3389/fnhum.2016.00223
12. Guillaume, C., Konstantina, K., Thierry, P.: GamEMO: how physiological signals show your emotions and enhance your game experience. In: Proceedings of the 14th ACM International Conference on Multimodal Interaction - ICMI 2012, pp. 297–298. ACM Press, New York, October 2012. https://doi.org/10.1145/2388676.2388738
13. Hochreiter, S., Schmidhuber, J.: Long short-term memory. Neural Comput. **9**(8), 1735–1780 (1997). https://doi.org/10.1162/neco.1997.9.8.1735
14. Holmgård, C., Yannakakis, G.N., Martinez, H.P., Karstoft, K.I.: To rank or to classify? Annotating stress for reliable PTSD profiling. In: 2015 International Conference on Affective Computing and Intelligent Interaction (ACII), pp. 719–725. IEEE (2015)
15. Kivikangas, J.M., et al.: A review of the use of psychophysiological methods in game research. J. Gaming Virtual Worlds **3**(3), 181–199 (2011). https://doi.org/10.1386/jgvw.3.3.181_1
16. Koster, R.: Theory of Fun for Game Design. O'Reilly Media, Inc., Sebastopol (2013)
17. Lang, P.J., Greenwald, M.K., Bradley, M.M., Hamm, A.O.: Looking at pictures: affective, facial, visceral, and behavioral reactions. Psychophysiology **30**(3), 261–273 (1993)
18. LeCun, Y., Bengio, Y., Hinton, G.: Deep learning. Nature **521**(7553), 436–444 (2015). https://doi.org/10.1038/nature14539
19. Liu, C., Agrawal, P., Sarkar, N., Chen, S.: Dynamic difficulty adjustment in computer games through real-time anxiety-based affective feedback. Int. J. Hum. Comput. Interact. **25**(6), 506–529 (2009)

20. Lopes, P., Liapis, A., Yannakakis, G.N.: Framing tension for game generation. In: Proceedings of the International Conference on Computational Creativity (2016)
21. Lopes, P., Yannakakis, G.N., Liapis, A.: RankTrace: relative and unbounded affect annotation. In: 2017 7th International Conference on Affective Computing and Intelligent Interaction (ACII), pp. 158–163. IEEE (2017)
22. Maier, M., Elsner, D., Marouane, C., Zehnle, M., Fuchs, C.: DeepFlow: detecting optimal user experience from physiological data using deep neural networks. In: Proceedings of the 28th International Joint Conference on Artificial Intelligence, vol. 2019, pp. 1415–1421. International Joint Conferences on Artificial Intelligence Organization, August 2019. https://doi.org/10.24963/ijcai.2019/196
23. Malik, J.: What led computer vision to deep learning? Commun. ACM **60**(6), 82–83 (2017). https://doi.org/10.1145/3065384
24. Martínez, H.P., Bengio, Y., Yannakakis, G.N.: Learning deep physiological models of affect. IEEE Comput. Intell. Mag. **8**(2), 20–33 (2013). https://doi.org/10.1109/MCI.2013.2247823
25. Perez Martínez, H., Garbarino, M., Yannakakis, G.N.: Generic physiological features as predictors of player experience. In: D'Mello, S., Graesser, A., Schuller, B., Martin, J.C. (eds.) ACII 2011. LNCS, vol. 6974, pp. 267–276. Springer, Heidelberg (2011). https://doi.org/10.1007/978-3-642-24600-5_30
26. Mühl, C., Allison, B., Nijholt, A., Chanel, G.: A survey of affective brain computer interfaces: principles, state-of-the-art, and challenges. Brain Comput. Interfaces **1**(2), 66–84 (2014). https://doi.org/10.1080/2326263X.2014.912881
27. Pagulayan, R.J., Keeker, K., Wixon, D., Romero, R.L., Fuller, T.: User-centered design in games. In: The Human-Computer Interaction Handbook: Fundamentals, Evolving Technologies and Emerging Applications, pp. 883–906. L. Erlbaum Associates Inc. (2002)
28. Poh, M.Z., Swenson, N.C., Picard, R.W.: A wearable sensor for unobtrusive, long-term assessment of electrodermal activity. IEEE Trans. Bio-Med. Eng. **57**(5), 1243–1252 (2010). https://doi.org/10.1109/TBME.2009.2038487
29. Rani, P., Sarkar, N., Liu, C.: Maintaining optimal challenge in computer games through real-time physiological feedback. In: 11th HCI International, Las Vegas, USA. Lawrence Erlbaum Associates Inc. (2005)
30. Schuller, B., Batliner, A., Steidl, S., Seppi, D.: Recognising realistic emotions and affect in speech: state of the art and lessons learnt from the first challenge. Speech Commun. **53**(9–10), 1062–1087 (2011). https://doi.org/10.1016/j.specom.2011.01.011
31. Sequeira, H., Hot, P., Silvert, L., Delplanque, S.: Electrical autonomic correlates of emotion. Int. J. Psychophysiol. **71**, 50–56 (2009)
32. Shaker, N., Yannakakis, G.N., Togelius, J.: Towards automatic personalized content generation for platform games. In: 6th Artificial Intelligence and Interactive Digital Entertainment Conference, AIIDE (2010)
33. Spronck, P., Ponsen, M., Sprinkhuizen-Kuyper, I., Postma, E.: Adaptive game AI with dynamic scripting. Mach. Learn. **63**(3), 217–248 (2006). https://doi.org/10.1007/s10994-006-6205-6
34. Trigeorgis, G., et al.: Adieu features? End-to-end speech emotion recognition using a deep convolutional recurrent network. In: 2016 IEEE International Conference on Acoustics, Speech and Signal Processing (ICASSP), pp. 5200–5204. IEEE March 2016. https://doi.org/10.1109/ICASSP.2016.7472669
35. Wang, C., Pun, T., Chanel, G.: A comparative survey of methods for remote heart rate detection from frontal face videos. Front. Bioeng. Biotechnol. **6**, 33 (2018). https://doi.org/10.3389/fbioe.2018.00033

36. Yannakakis, G.N., Martínez, H.P., Jhala, A.: Towards effective camera control in games. User Model. User-Adap. Interact. **20**(4), 313–340 (2010). https://doi.org/10.1007/s11257-010-9078-0
37. Zheng, W.l., Zhang, Y.-Q., Zhu, J.Y., Lu, B.l.: Transfer components between subjects for EEG-based emotion recognition. In: 2015 International Conference on Affective Computing and Intelligent Interaction (ACII). pp. 917–922. IEEE, September 2015. https://doi.org/10.1109/ACII.2015.7344684

Understanding Challenges Presented Using Emojis as a Form of Augmented Communication

Mariam Doliashvili, Michael-Brian C. Ogawa, and Martha E. Crosby[✉]

University of Hawai'i at Mānoa, Honolulu, HI 96822, USA
{mariamd,ogawam,crosby}@hawaii.edu

Abstract. Emojis are frequently used in social media and textual conversations. They are significant means of communication to visually help express emotions and describe objects. Previous studies have shown positive impacts of emojis used in human relations, memorization tasks and engagement with web content. Unicode version 6 includes 2923 emojis, which makes it difficult to use them effectively without a recommender system. We formulate recommending emojis as a complex prediction problem based on its diverse usage as a word and as a sentiment marker. People have individual usage and representations of emojis across different platforms as well as different interpretations based on the device. In order to increase the accuracy of the recommending process, we propose using a recommender system that applies personalization to suggest various emojis. This paper describes several baseline models we explored and the Long Short-Term Memory (LSTM) recurrent neural networks we trained to test the feasibility of extracting knowledge from emoji datasets to predict emoji usage.

Keywords: Augmented communication · Recommender system · Emoji usage

1 Introduction

1.1 Relevance

Emojis are widely used on social media sites like Twitter, Facebook and Instagram to enrich the conversation with emotions and elaborate meaning in fewer words. Their use, however, is not limited to social media: Chat groups, mail and text messages are examples of other areas where emojis are employed. Emojis are a quick and easy to use way to express one's emotions in online communications. They appeared on Japanese mobile phones in the 1990s. The origin of the word emoji [25] comes from Japanese e (絵, "picture") + moji (文字, "character"). Their popularity has increased internationally in the past two decades. Many people prefer texts with emojis and according to an AMEX OPEN Forum infographic [3], emojis can make a big difference to "post" engagement rates. Posts with emojis get 33% more comments, they are shared 33% more often and they get liked 57% more often than posts without them. A study of the traits of highly influential social media profiles by Simo Tchokni et al. [21] showed the use of emoticons was common factor among these powerful users. However, there are specific cases of emoji usage where emoji usage does not give positive outcomes.

© Springer Nature Switzerland AG 2020
D. D. Schmorrow and C. M. Fidopiastis (Eds.): HCII 2020, LNAI 12196, pp. 24–39, 2020.
https://doi.org/10.1007/978-3-030-50353-6_2

1.2 Relevance

It is somewhat ambiguous, whether one should use an emoji at work related emails or not. Jina Yoo [28] tested how people perceive smiley faces in work email as compared to social email. Researchers sent two types of email messages to a group: a flirtatious message, and another one about extending a job interview request. Emoticons were added to some texts of each type. "The usage of emoticons induced stronger relational outcomes in a task-oriented context than in a socioemotional context. Participants in a task-oriented condition liked the sender more if the sender used an emoticon rather than if the sender used no emoticons" and the sender's credibility wasn't affected by the emoticons even when they used up to four emoticons. As the possible explanation of the result is given the following: "emoticons are overused already in socio-emotional contexts, and no special value is assigned to using emoticons in email in the same context. However, when the emoticons are used in a task-oriented context, they might function as a positive expectancy violation, which could bring positive relational outcomes." Contrary to this, Glikson et al. [14] published a paper "The Dark Side of a Smiley, Effects of Smiling Emoticons on Virtual First Impressions" where it is stated that "contrary to actual smiles, smileys do not increase perceptions of warmth and actually decrease perceptions of competence" and "Perceptions of low competence, in turn, undermined information sharing." (The authors also mentioned that if all the team members were younger, the likelihood of using emoticons in the team's conversation was higher). We cannot draw a general conclusion as both studies are evaluated in specific scenarios (1 - extending a job interview request, 2 - First impressions over internet). Most of these papers study only emoticons not visual emojis as emojis have only been commonly available since 2010.

Wang et al. [26] showed that emoticons reduced the negativity effect in business-related email messages, that is the same message sounded less negative when paired with a positive (smiley) emoticon. In addition, emojis have been shown to lead to better memorization of content [10]. Kalyanaraman et al. [17] conducted a study that had participants chat online with "health experts" and "film experts" who either used or avoided emoticons, the participants rated the experts in both topics friendlier and more competent when they communicated with emoticons. This study also noted that emoticons might help you remember what you've read – "It appears that the presence of emoticons affects cognition as well, because participants" scores on memory for chat content were significantly higher in the "emoticons present" condition than in the "emoticons absent" condition."

1.3 History

Emojis are often confused with emoticons. An emoticon is a representation of a facial expression using punctuation marks, numbers and letters, usually written to express a person's feelings or mood, e.g. While emojis are used like emoticons they are small digital images or icons that exist in various genres, including facial expressions, common objects, food, activities, animals, places and types of weather. For example:

The first emoji was created in 1999 in Japan by Shigetaka Kurita [4]. However, "The development of emoji was predated by text-based emoticons, as well as graphical representations, inside and outside of Japan. Scott Fahlman, a computer scientist at Carnegie Mellon University, was credited with popularizing early text-based emoticons in 1982 when he proposed using the following: -) and: - (sequence of characters to help readers of a school message board distinguish between serious posts and jokes. From 2010 onwards, hundreds of emoji character sets have been incorporated into Unicode, a standard system for indexing characters, which has allowed them to be used outside Japan and to be standardized across different operating systems. After 2010 each update of the Unicode standards introduced new sets of emojis. "Corporate demand for emoji standardization has placed pressures on the Unicode Consortium, with some members complaining that it had overtaken the group's traditional focus on standardizing characters used for minority languages and transcribing historical records [4]."

2 Difficulties of Building an Emoji Recommender System

The fact that emojis are related to emotions and are becoming means of communication makes emoji prediction an interesting problem for Natural Language Processing (NLP). If we assume that an emoji is a label of the text corresponding to an emotion, then we would face the sentiment analysis problem. However, the classical sentiment analysis problem only analyses whether a text is positive or negative – sentiment polarities of sentences and documents. Advanced models only have several additional emotions like happiness and anger. On the other hand, emoji classification has a larger population of candidates. As of November 2017 - there are 2623 emojis available in unicode [16]. They are much more detailed and complicated to predict, because one emoji corresponds to many emotions based on its use and the same emotion can be expressed with various emojis. Sentiment analysis using emojis as emotional markers would be a tedious task. Emojis not only express emotions, they also describe professions, activities, flags, food and have gender and racial diversity. This diversity and quantity of emojis makes a recommender system necessary.

Furthermore, recommending emojis in a chat application also requires the understanding of a conversation, since an emoji can be used as:

- An answer to the previous text if it was a question,
- A reply for the previous text,
- A next word in the sentence.

If we were to fully cover the task of emoji prediction, we would need to address question-answering, smart reply and next word prediction problems. Next word prediction is covered because it is hard to determine a start and an endpoint of a conversation. People do not necessarily use the same emojis in the same situations, they might have completely different emoji usage and conversation styles. For example, consider the conversation of two texts: text 1 from Bob to Alice and text 2 from Alice to Bob. Text 1 is lengthy and includes emojis that are mostly negative, however, Alice never uses negative emojis.

3 Description of the Task

The problem can be stated as: Predict the top K emojis that Alice would use after writing a short message that may or may not already include emojis. The Facebook sentiment analysis paper by Tial et al. [23] shows that a sentence with an emoji does not necessarily equal to the same sentence without emojis. Thus, both the words in a sentence and the emojis in the text need to be considered. Studying the Twitter dataset shows three main cases of emoji usage with a text:

- Expressing an emotion about the text it accompanies.
  i.e. "Last couple months have been crazy! 😵😵 "
- As a word in the sentence.
  i.e. "I 💜 you", "Damn, I would love this. Or suicide squad, working towards that.
  Patty was 🔥💯 "
- Emoji combinations to express…

  - … a phrase – " ☕🕐 ?"- what time should we get coffee?. "🚗📧➕ ?"- Do you need a ride to the airport? "Had an accounting midterm today that I wasn't expecting until next week. Lots of unfamiliar terms 😵🔫. Here 😵🔫 stands for 'kill me now'.
  - … the strength of emotion – "😿😿😿"- very sad. "👏   👏   👏👏👏"- bravo. 😂😂😂 - very funny. etc.
  - … an emotion/entity that has no separate emoji in the unicode yet? With me now: Every damn corner 🌮🚚 #TacoTuesday? Here 🌮🚚 stands for a taco food truck. "🌱🐛🦋🐛🌱"- life cycle. "Anti-gravity 🏢🚶 " - 🏢🚶 stands for 'city parkour '.

The given sentence is studied for the above aspects, but there are many other factors besides the text that affect the usage of emoji. The relations between emojis and emotions are ambiguous, given that there are many emojis [12] to express the same emotion and the same emoji can be used to express different emotions [18]. Furthermore, the same emoji can have very different meanings for different people, and it can lead to misunderstandings [24]. This becomes an obvious problem by looking at the different representations of the same unicode emoji across various platforms. Miller et al. [18] shows differences in the way the same person perceives the same emojis. People can change their emoji usage while typing, based on the device they use.

Various studies over the years show that emoji usage can be different based on gender (females are using emojis with tears more frequently [2]) and culture (Different countries have different favorite emojis [1]). Emoji usage can also depend on the location of the person texting. For instance, a person in Hawaii is more likely to use palm tree and pineapple than a person in Siberia.

Emoji usage can also depend on the weather in the area. For example, if Hawaiian residents visit Mauna Kea (a dormant volcano whose peak is the highest point in the state of Hawaii) in winter, they are likely to use winter emojis. The location and nationality of the person can predict the flags that are used. The dataset suggests that people in the same friend circle tend to use similar emojis and vocabulary. People also tend to use more emojis with short texts than with larger texts. In addition, age can be an interesting feature to observe, not only because there are differently aged emojis, but also people in different generations view emojis differently. These differences make it very difficult to make individual predictions from a general dataset, so the decision was made to personalize the training and create a system that recommends emojis. Taking age, gender, location, and culture into account, it is possible to create user profiles and combine similar profiles for training and prediction.

3.1 Contribution

Unlike existent models that limit the existing unicode set of emojis to a smaller subset, we created a recommender system to personalize emoji predictions. This system categorizes the emoji prediction problem as 'sentiment analysis', 'next word prediction' and 'word-emotion mapping' tasks, and combines these three models with a heuristic to recommend the top k most relevant emojis.

4   Related Work

Related work on emoji predictions focus on emojis as a sentiment marker. A recent study from Barbieri et al. [5] addresses the prediction problem using a multi class LSTM classifier to predict one emoji per tweet. They mentioned that their system predictions are more accurate than human evaluators, since it is not easy for humans to predict emoji from the text without background knowledge about the text and the person. This combined with the fact that individuals have a different understanding of the same emojis [18] suggests that predictions are dependent on individual people and their way of writing and expressing emotions. In addition, the same study [18] shows that because of the different representations of emojis across different platforms people assign various emotions to the same unicode emoji. Therefore, our system to personalize emoji recommendations includes "individual device" as a separate entity. Sentiment analysis is an interesting research topic for NLP, however, there are not many studies about emoji usage and predictions. Barbieri et al. [5] address the problem to predict one emoji per tweet but they only attempt to predict X different emojis. Currently, there is no research that analyses data using all 2623 emojis. Papers concerning emoji predictions [5, 27] limit the number of emojis used to under 65 of the most popular ones. Most of the models with good performance (accuracy >50%) [5, 27] classify under 10 emojis. Since some of the platforms, sort the emoji list putting the most used ones on the top, recommending only one of the top k emojis does not help the users. User satisfaction and emoji popularity could be increased if relevant to the text emojis were recommended even though they were infrequently used and hard to find.

5   Dataset and Methods

For this study, four datasets were collected from Twitter, including 600,000 tweets (Dataset #1) and 50,000 tweets (Dataset #2) for the following 74 emojis :

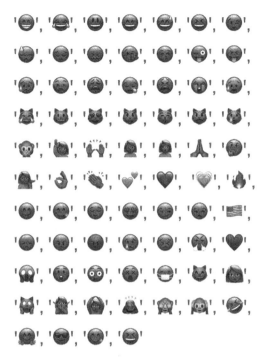

Fig. 1. 74 popular emojis on Twitter that are used to create the datasets #1 and #2.

The third dataset (Dataset #3) was collected for a limited amount of emojis and amounts to 50,000 tweets:

Fig. 2. 10 popular emojis on Twitter that were used to create dataset #3.

Complete datasets gathered from five volunteers were used to study the personalized emoji recommendations.

5.1 Methods

The following formulation of tasks was designed for the purpose of this study:

- The initial goal is to predict an emoji used in the tweet, in any possible position, based on the general dataset (the tweets are gathered from multiple users in a 5 to 10 h time frame).
- Considering the complexity and diversity of emoji prediction, the problem was simplified by making it a binary classification problem: For a chosen emoji, the prediction was whether it will be used in the tweet.
- Finally, predict an emoji that is used after the given text, based on one user's data.

- The fact that one emotion can be described with various emojis suggest that for better results we should recommend a set of emojis that correspond to the same emotion. With this modification the binary classification problem changes to predict whether a tweet contains an emoji from the given set and for the general prediction we can recommend top k emojis based on their probabilities.

5.2 Dataset

For not personalized predictions the data was collected using the Twitter Streaming API [11] in the time period of April 29–May 1, 2017. The language used in the dataset is English. The only preprocessing that has modified the dataset was removing the unnecessary information: URLs, #-signs, malformed words containing numbers, etc. However, the data contains a significant amount of spelling errors so using spelling correction might increase the quality of the dataset and lead to a better solution.

The above datasets are labeled in the two different ways:

- Each tweet labeled with the emoji used in it.
- Each tweet labeled with 1/0 based on whether a specific emoji/or an emoji from a specific set of emojis is used in the tweet.

The second case also needs balancing the dataset after the labeling, since it can result in a disproportionate amount of data for a binary classification problem. One more thing to mention is that there are many cases when a tweet contains several emojis. In this case, we labeled the tweet with the first occurring emoji. The emoji is by default one of the labels of the tweet and the choice of the first emoji is random; in addition labeling the same tweet with various labels would create a problem that we would have to address later on the learning phase.

For personalized prediction the data for training, testing and evaluation is from the same Twitter user and includes 47000 tweets. For labeling the dataset we use all available unicode emojis. We created a mapping of each emoji to its description and key words that it associates with. Tweets are split into sentences that are followed by an emoji and then labeled with the emoji that was following the text.

E.g. I like that <3 thought I would not participate :/
Would produce the following sentences and labels:
Sentence 1: I like that
Label 1: <3
Sentence 2: I like that <3 thought I would not participate
Label 2: :/

In addition, we took the combinations of emojis that were used together to express a non-existent emoji or a phrase (in short we call them combojis) into account and added them into the labels' list. For each label we calculate and update the frequency of how many times it is used. When combojis frequency reaches to certain limit we generate a mapping and keywords for it, based on previous use. The frequency is also used as a feature for training.

5.3 Word Representations

For the word representations, we used one hot encoding and word embeddings.

- **One Hot Encoding.** For one hot encoding [8], we calculate the frequency of words in all tweets. Then, we take the k most frequently used words and create a binary vector for each tweet. Each binary vector has 5,000 entries, where an entry corresponds to one of the k most frequent words. Entries are filled with an 1, if the corresponding word is present in the tweet and 0 if it is not.
- **Word Embeddings.** For word embeddings [6], we, again, calculate the frequency of words in all tweets. For each tweet, we create a vector with as many entries as there are words in a tweet. Each entry is filled by using a word's index in the k most frequently used words as the value. If a word does not occur in the k most frequent words, we fill the vector entry with zero. In the end, we train the word embeddings with the neural network. Similar words should be placed close together in the vector space after the learning. Finally, the resulting word representations are split up into training, testing and evaluation sets.

5.4 Personalized Learning

The model for personalized learning combines three solutions for subtasks and a scoring function. Emojis and combojis can be expressing an emotion and a next word in the sentence. Therefore, we combined the approaches of next word suggestion, algorithm and sentiment analysis.

5.5 Next Word Suggestion Algorithm

We need to construct an algorithm that fulfills the following steps: Build a language model using twitter text and then use this language model to predict the next word as the user types.

We need to calculate the frequency of words and n-grams and use a sliding window.

If we assume the training data shows the frequency of "university" is 198, "university student" is 12 and "university professor" is 10. We calculate the maximum likelihood estimate (MLE) as:

- The probability of "university student":
  $Pmle\ (entry|data) = 12/198 = 0.06 = 6\%$
- The probability of "university professor" is:
  $Pmle\ (streams|data) = 10/198 = 0.05 = 5\%$

If the user types, "university", the model predicts that "student" is the most likely next word.

The n-gram model description steps are:

- Generate 2-grams and 3-grams.
- Select n-grams that account for 60% of word instances. This reduces the size of the models.
- Calculate the maximum likelihood estimate (MLE) for words, for each model.

As for the prediction: We use the ngram models on tokenized and preprocessed user input. We implement a Stupid backoff [7] starting on the 3-gram model backing off to the 2-gram model and returning 3 words with the largest MLE.

5.6 Mapping from a Word to Emoji

We created the mapping using 2623 available unicode emojis from unicode.org and their descriptions as names. We generate keywords based on the frequency of the words that they are used with in a sentence. The mapping file is updated to store information for combojis. The file will be used to map the next predicted word to associated emojis.

5.7 Evaluating Emotion

Both in general and personalized tasks we train a Long Short-Term Memory (LSTM) network to predict which emojis are used with the text. LSTM is a special kind of Recurrent Neural Network (RNN) able to learn long term dependencies. This kind of RNNs are good at remembering information for long periods of time.

LSTMs were introduced by Hochreiter and Schmidhuber [15]. Over the course of years they were popularized and developed by various contributors, since they perform well on a large variety of problems. Over the course of years, LSTMs proved to perform well for NLP tasks including sentiment analysis.

5.8 Scoring

Scoring of emojis in the final stage is based on the prediction probability produced by a trained model for labeling, label frequency in the existing dataset that the model is trained on and an additional feature for measuring confidence.

Out of the above two models we get two sets of predictions for each emoji. The next word prediction task with mapping assigns probabilities to the possible labels (emojis) - whether they have been used before or not. The sentiment prediction part uses already used emojis as labels. Therefore, it generates probabilities for a limited set of emojis. We have a confidence score to adjust the two probabilities based on the relative frequency of emojis.

For example: If an emoji has a prediction probability 0.9 and its relative frequency is in the top 10%, we give it a high confidence score. Finally, we calculate the weighted sum of the prediction probability and confidence and recommend the top k emojis for a given text.

5.9 Implementation

While achieving the results of this paper required a substantial amount of scripting, three Python libraries were essential to the analysis.

- **NLTK [19].** The Natural Language ToolKit (NLTK) is a Python library specialized in natural language processing. We made use of its word tokenizing capabilities and used its naive Bayes classifier to create the baseline method.
- **SciKit [13]** Scikit is a machine learning library for python that offers tools for data mining and analysis. We have used its implementations of the algorithms - Logistic Regression and Stochastic Gradient Descent.
- **Keras on TensorFlow [20, 22].** Keras is a high-level neural networks API, written in Python and capable of running on top of either TensorFlow or Theano. Tensorflow, an open-source library for numerical computations, was developed by Google Brain Team for the purposes of conducting machine learning and deep neural networks research. This study used it to train word embedding vectors, build an LSTM classifier and optimize it for accuracy.

The NLTK and SciKit classifiers are used with the one hot encoding of the tweets. Each tweet is labeled with the emoji, that is contained in the tweet.

The LSTM neural network is used with word embeddings vectors. The dataset has been labeled in the following ways:

- For a specific emoji:

  1. The label for each tweet equals to 1 if the tweet contains the emoji.
  2. Otherwise the label equals to 0.

- For a specific emotion:

  1. Create a set of emojis that correspond to the emotion.
  2. The label for each tweet equals to 1 if the tweet contains an emoji from the given set.
  3. Otherwise the label equals to 0.

The emojis used for a single emoji classifier are:

The emoji sets used related to the same emotion are:
The example of combojis generated after training on an individual data are:

6 Evaluation Methods

For the general task we use the prediction accuracy as a metric.

Fig. 3. List of the emojis grouped by relation to similar emotions.

Fig. 4. Combojis that were generated after a training process on the individual data do-nated by a twitter user.

As for the individual prediction we have two evaluation metrics: Precision at the top k candidates and the mean reciprocal rank that is used to evaluate the ranking of the top predictions.

In the future, it is necessary to use human evaluation - If people did not use emojis in a text that does not mean that they would not use them if a recommender system was available.

7 Analysis

Since the computations used in the implementation are quite time costly, the classifiers for multi-variable problem were trained on the smaller dataset (Dataset #2). For the results please see the Table 1.

Table 1. Accuracy of the naive approach

Dataset #2			
Algorithm	5 emojis	10 emojis	20 + emojis
Naive Bayes	48%	39%	32%
Logistic regression	52%	38%	32%
Stochastic gradient descent	13%	9%	7%

Unfortunately, Stochastic Gradient Descent was not a fit for the problem, as for the Naive Bayes and Logistic Regression they have shown improvements when hyper-parameters were reset. The LSTM neural network was implemented solely for the binary classification problems, using word embeddings. It has been directed to optimize the accuracy over the training phase. This resulted in an average of 71% accuracy for a single emoji classifier, and an average of 70% accuracy for the classifier of a set containing 4 emojis related to the same emotion. The emoji sets related to the same emotion were chosen naively and for the future improvements it is necessary to explore the ways to create the emojis sets that are the most related to each other and interchangeable in the everyday usage.

The following figure summarizes the overall evaluation for the average values of accuracy:

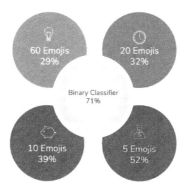

Fig. 5. Accuracy measures for different set of emojis.

The binary classifier result demonstrated in the figure is trained, tested and evaluated on 6:2:2 proportions of 365,000 tweets. The original dataset used is Dataset #1, which after preprocessing and balancing is reduced to 365,000 tweets. A binary classifier is used to determine for each emoji whether it should be recommended for a tweet or not. Given the result we can assume that, if recommendation for an emoji is given with 71% of accuracy, a recommender system that uses suggestion of three top results would significantly increase overall accuracy of the suggestion.

Experiments [5] showed that human evaluators on average achieve 80% accuracy on twitter dataset. Sine the accuracy of the recommendation for each emoji equals to 0.71, the system is close to human performance.

To compare the results with existing papers, we created a combination of 5 binary classifiers that predicts the one emoji that has the highest probability out of 5. However we need to take into account that the datasets are not the same. Especially Xie et al. [27] who use Weibo data in Chinese language.

Accuracy for the next word suggestion algorithm using stupid backoff is 13.5%. After mapping the words with emojis and combining it with LSTM predicted emojis recommender system had the accuracy of 34% on average. While LSTM only achieves 74%.

In fact, incorporating the Next Word Prediction (NWP) algorithm may decrease the performance measure. Since NWP together with mapping does not limit any emojis from the full unicode emoji list [16] it forms recommendations by including a wide

Evaluating Precision

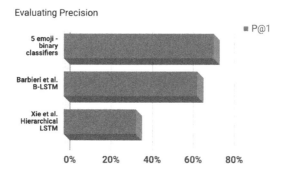

Fig. 6. P@1 evaluation for 5 emoji binary classifier compared to other papers.

Evaluating Precision

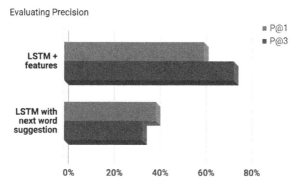

Fig. 7. P@1 and P@3 evaluations for LSTM recommenders.

range of rarely used emojis. The users in our personalized dataset only chose up to 125 emojis out of the 2623 available. In order to better evaluate the accuracy of our recommender system, we need active Twitter users to train on their dataset and record useful recommendations in real-world setting.

8 Conclusion

Emojis are used in everyday life by millions of people, which makes them a widely used tool for expressing emotions. A vast amount of data is available for experimenting with Machine Learning algorithms for an emoji related problem. It is an interesting NLP task to analyze and explore the ways they relate to emotions. In addition, the existence of emojis gives us the opportunity to research the emotions of online usage of particular events such as analyzing the sentiment towards U.S. Presidential Candidates [9]. Potential applications give software developers the incentive to create systems that make it easier for users to include emojis in their texts [29]. This paper shows that it is possible to create a recommender system for emojis with reasonable accuracy.

8.1 Multi-variable vs. Binary Classifier

The task of creating a multi-variable classifier is easy to formulate. It is also easy to create the necessary dataset. However, it required exploring emoji usage first: Which emojis correspond to the same emotion? How many emojis to select for a single study? Which emojis are used together? It is possible to predict usage of one emoji. Solving the problem for individual cases lets us formulate the solution of the initial problem. Recommending several emojis from the same classifier also increases the accuracy of a recommender system.

Additionally, there are also issues to address regarding the binary classifier. It introduces bias and needs balancing of the dataset.

Another way to resolve all the differences in emoji usage for a multi classifier is personalizing the dataset for a particular user. This helps develop a system that would be highly useful for active users, but can be problematic for the users that do not have enough data for the model to learn. Finally, generalizing task by not restricting the dataset but creating user profiles can have a potential to achieve a good result. Since it would be possible to create a personalization element for ambiguous emojis and benefit from the knowledge acquired from a larger dataset.

8.2 Future Work

As future work, we can do improvements for various steps of the development:

- **Dataset**:
  - Training the model on a bigger dataset from individual users to make it possible to profile based on features like location and ethnicity.
  - We need to evaluate accuracy of the above methods compared to human operators.

- **Clustering emojis**: Exploring ways to create improved mapping (emoji:words) for the classifiers:
  - Set of emotions that have corresponding emojis without intersection.
  - Set of the emojis that correspond to the same emotion.

The latter would allow us to improve the error function, since recommending an emoji related to the correct label should be weighted as a smaller error. Ultimately the system performance must be tested in a real-world setting.

Acknowledgements. Special thanks to Dr. Lipyeow Lim and Dr. David Chin for useful advices. Ed White and Gene Park for collecting and donating personal data. [11–13, 19, 20, 22]

References

1. Swiftkey emoji report (2015). https://www.scribd.com/doc/262594751/. SwiftKey-Emoji-Report. Accessed 4 Dec 2017

2. Emoji gender study (2017). https://www.brandwatch.com/blog/react-gender-emoji-data/. Accessed 4 Dec 2017
3. Emojis effect on facebook engagement (2017). https://blog.bufferapp.com/7-facebook-stats-you-should-\know-for-a-more-engaging-page. Accessed 4 Dec 2017
4. History of emojis (2017). https://en.wikipedia.org/wiki/Emoji#History. 4 Dec 2017
5. Barbieri, F., Ballesteros, M., Saggion, H.: Are emojis predictable? CoRR, abs/1702.07285 (2017)
6. Bengio, Y., Ducharme, R., Vincent, P., Janvin, C.: A neural probabilistic language model. J. Mach. Learn. Res. **3**, 1137–1155 (2003)
7. Brants, T., Popat, A., Xu, P., Och, F.J., Dean, J.: Large language models in machine translation, January 2007
8. Cassel, M., Kastensmidt, F.L.: Evaluating one-hot encoding finite state machines for SEU reliability in SRAM-based FPGAs. In: Proceedings of the 12th IEEE International Symposium on On-Line Testing, IOLTS 2006, Washington, DC, USA, pp. 139–144. IEEE Computer Society (2006)
9. Chin, D., Zappone, A., Zhaon, J.: Analyzing Twitter sentiment of the 2016 presidential candidates (2015)
10. Danesi, M.: The Semiotics of Emoji: The Rise of Visual Language in the Age of the Internet. Bloomsbury Advances in Semiotics. Bloomsbury Publishing, London (2016)
11. Twitter Developers. Twitter streaming API (2017). https://dev.twitter.com/streaming/overview. Accessed 4 Dec 2017
12. Emojipedia. Emojis available for apple products (2017). https://dev.twitter.com/streaming/overview
13. French Institute for Research in Computer Science and Automation (INRIA). Scikit (2017). http://scikit-learn.org/. Accessed 4 Dec 2017
14. Glikson, E., Cheshin, A., van Kleef, G.A.: The dark side of a smiley: effects of smiling emoticons on virtual first impressions. Soc. Psychol. Personal. Sci. **9**, 614–625 (2018)
15. Hochreiter, S., Schmidhuber, J.: Long short-term memory. Neural Comput. **9**(8), 1735–1780 (1997)
16. Unicode Inc. Unicode version 5 emojis full list (2017). https://unicode.org/emoji/charts-5.0/full-emoji-list.html. Accessed 4 Dec 2017
17. Kalyanaraman, S., Ivory, J.D.: The face of online information processing: effects of emoticons on impression formation, affect, and cognition in chat transcripts, 17 December 2013
18. Miller, H., Thebault-Spieker, J., Chang, S., Terveen, L., Hecht, B.: "Blissfully Happy" or "Ready to Fight": Varying Interpretations of Emoji, pp. 259–268. AAAI press, Hilversum (2016)
19. Team NLTK. Natural language toolkit (2017). http://www.nltk.org/. Accessed 4 Dec 2017
20. Chollet, F., et al.: Keras: the Python deep learning library (2017). https://keras.io/. Accessed 4 Dec 2017
21. Tchokni, S., Seaghdha, D.O., Quercia, D.: Emoticons and phrases: status symbols in social media (2014)
22. Google Brain Team. Tensorflow (2017). https://www.tensorflow.org/. Accessed 4 Dec 2017
23. Tian, Y., Galery, T., Dulcinati, G., Molimpakis, E., Sun, C.: Facebook sentiment: reactions and emojis, January 2017
24. Tigwell, G.W., Flatla, D.R.: Oh that's what you meant!: reducing emoji misunderstanding. In: Proceedings of the 18th International Conference on Human-Computer Interaction with Mobile Devices and Services Adjunct, MobileHCI 2016, pp. 859–866. ACM, New York (2016)
25. Cambridge University. Emoji meaning in the Cambridge English dictionary (2017). http://dictionary.cambridge.org/dictionary/english/emoji. Accessed 4 Dec 2017

26. Wang, W., Zhao, Y., Qiu, L., Zhu, Y.: Effects of emoticons on the acceptance of negative feedback in computer-mediated communication. J. Assoc. Inf. Syst. **15**, 454–483 (2014)
27. Xie, R., Liu, Z., Yan, R., Sun, M.: Neural emoji recommendation in dialogue systems. CoRR, abs/1612.04609 (2016)
28. Yoo, J.H.: To smile or not to smile :): defining the effects of emoticons on relational outcomes, 24 May 2007
29. Zhou, H., Huang, M., Zhang, T., Zhu, X., Liu, B.: Emotional chatting machine: emotional conversation generation with internal and external memory. CoRR, abs/1704.01074 (2017)

Investigation of Biological Signals Under the Stimulation of Basic Tastes

Masaki Hayashi, Peeraya Sripian[(✉)], Uma Maheswari Rajagopalan, Runqing Zhang, and Midori Sugaya

Shibaura Institute of Technology, 3-7-5, Toyosu, Koto-ku, Tokyo 135-8504, Japan
{doly,peeraya}@shibaura-it.ac.jp

Abstract. Recently, neuromarketing has attracted attention as it is a marketing technique analyze consumer's emotion using biological signal, which could genuinely understand consumers. From recent studies using the observation of brain wave (EEG), there are changes in emotional patterns (among the four emotions: Happy, Angry, Sad, Relax) according to the satisfaction on taste. In this work, we investigate the biological signals (EEG and ECG) under the stimulation of basic tastes: sweetness, saltiness, sourness, bitterness and savory taste (umami), while taking into account individual preference in the evaluation. In the experiment, all subjects were test cognitive detection threshold before proceeding to the main experiment. The level of concentration that yields a just-noticeable threshold taste was used as the cognitive threshold for each subject for EEG and ECG data collection. The biological signals were used to classify emotion and compare with subjective evaluation. The experiment results proved that human emotion is effected by the taste stimuli.

Keyword: Taste · Biological signals · Emotion classification · Neuromarketing · Consumer behavior

1 Introduction

At present, conventional marketing methods such as questionnaires, survey paper, experiment, observation, purchase history data, and so on cannot understand consumers' needs properly [1]. Recently, neuromarketing, the application of neuroimaging methods to product marketing has attracted attention as it analyzes biological signal of the consumer. Neuroimaging is promising to become cheaper and faster than other marketing methods and it could provide marketers with information that could not be obtained using conventional marketing methods [2]. Neuroimaging methods promises possible marketing in precedence of a product even before the release, i.e., almost at the stage of conception. Simpler approaches such as focus groups, survey could work as tools of low cost but would include biases. On the other hand, market tests could provide accurate results at a higher cost [3].

One promising category for the application of neuromarketing is to food products as it would be flavor, taste, texture and aesthetic values and so on that could involve

© Springer Nature Switzerland AG 2020
D. D. Schmorrow and C. M. Fidopiastis (Eds.): HCII 2020, LNAI 12196, pp. 40–49, 2020.
https://doi.org/10.1007/978-3-030-50353-6_3

multiple perceptions [4]. A few studies that employ MRI scanner have been conducted with simple sugar solutions to chocolate, wine, sports drinks and colas to investigate gustatory experience [5, 6]. Especially, in neuromarketing, investigation related to taste and the corresponding neurological signals is still limited or under-researched. Here, we are interested in the investigation of biological signals that include neurological ones under the stimulation of basic tastes.

Sugimoto et al. [7] analyzed the effects of gustatory stimulation on heart rate and electroencephalogram or brain waves. They found that stimulation umami taste activates the parasympathetic nervous system, while bitter taste stimulation activates the sympathetic nervous system. Based on EEG observations, they found changes among the four emotions: Happy, Angry, Sad, and Relax, depending on the perceived taste. While sweetness and sourness increased satisfaction, bitterness increased stress and decreased relaxation [7]. However, their research evaluated only the five basic tastes and did not perform Kansei evaluation such as individual preferences or the analysis of Kansei, and there was also no investigation done for the combination of tastes.

2 Proposal

The purpose of this study is to investigate the heart beat related EEG and brain activation related ECC signals under the stimulation of five basic tastes. As this is a pilot experiment toward the intended future investigation of the physiological signals under combination of tastes, we have conducted with restricted number of only three subjects. To conduct such an investigation under combination of different tastes, it would be necessary also to understand the preference of the individual and then incorporating such individual preferences into the evaluation.

3 Experiment

Table 1. Liquid stimuli used in the experiment

Liquid name	Mol concentration				
	1	2	3	4	5
Salt water (saltiness)	7.4	22.2	66.7	200.2	600.6
Caffeine (bitterness)	0.4	1.3	3.9	11.6	34.8
Sucrose (sweetness)	3.9	11.7	35.1	105.2	315.5
Citric acid (sourness)	0.1	0.3	0.8	2.3	7
Glutamic sodium (umami)	1	3	8.9	26.6	79.8

In order to evaluate the gustatory stimulation to a combination of tastes, we performed the preliminary experiment under five basic tastes, namely, sweetness, saltiness, sourness, bitterness, and umami/savory. In an earlier work conducted by a Japanese team, the study was done to classify emotion under different tastes by monitoring only the heart beats, i.e., through EEG. In this work, we extended the study to employ ECG in addition to EEG for evaluating to classify emotion. We expected such an approach would provide results that are more associated with the changes in emotion due to change in tastes. In classification of emotion, the emotion estimation model by Kawakami et al. [8] was used.

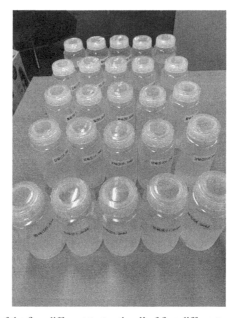

Fig. 1. A photograph of the five different taste stimuli of five different concentrations used in the experiment.

3.1 Subjects

Three male healthy subjects in the age range around 2 participated in the study. All the subjects were in normal healthy condition and did not have any taste related abnormalities. Written consents had been obtained before the start of the experiments. The subjects were requested not to have any food except water before the experiments.

The study was performed after getting the approval from the ethics committee of the university. Written consents were also obtained from the subjects before the starting the experiment and also they were told to withdraw at any time from the experiment. The study was conducted to test the feasibility of the whole experimental paradigm and so the study was restricted to three subjects. Subjects were given instructions about the experimental procedure and also they were fitted with EEG and ECG devices to understand the procedure of data collection. They were also assured that their personal details would be kept confidential.

Fig. 2. A photograph of the experiment. The subject wore a brain wave sensor and a pulse sensor during the whole experiment.

3.2 Stimuli

Five basic tastes namely, sweetness, saltiness, sourness, bitterness, and savory taste were used. For each of the taste, five different levels of diluted taste solutions were prepared and thus a total of 25 different solutions were prepared. Water, free of any taste, was used as the control. Table 1 shows the different tastes and their respective five levels of dilutions as Mol concentration used in the experiment. Figure 1 shows a photograph of the five different taste stimuli at five different concentrations used in the experiment.

3.3 Procedure

Cognitive Detection Threshold Testing

Before collecting the physiological signals, EEG and ECG, the threshold limit for perceiving of the each of the five tastes for the three different subjects was done. This was done through starting from the least dilution of the taste and then increasing till reaching the detectable limit.

The paradigm used in the experiment is given below in Fig. 3. The stimulus paradigm consists of 180 s of no stimulus followed by 60 s of water, followed by a period of 180 s when data was collected that included the resting period too. The subject was instructed first to taste the water solution mixed with the lowest taste concentration, and then the concentration was gradually increased in five levels until he/she can perceive the taste correctly. The level of concentration that yields a just-noticeable threshold taste was used as the cognitive threshold for each subject for EEG and ECG data collection.

3.4 EEG and ECG Data Collection

Once the threshold taste level for each of the participant was determined, physiological recordings of EEG and ECG were started. At first, the participant was asked to wear a wearable brain ECC recording device (Neurosky Ltd.) and a pulse sensor during the whole experiment (Fig. 2). The experimental procedure is as follows:

1. Subject stayed still (Rest) for 180 s for baseline measurement;
2. Subject was instructed to have the stimulus taste solution (5 ml) in the mouth without swallowing for the next 60 s;
3. After that, the subject was instructed to spit the solution out in a designated container, followed by water gargling.
4. In the next 180 s, subject first answered the questionnaire regarding the taste of the liquid and remained still (Rest) for the rest of the time.
5. 1–4 was repeated for all the stimuli.

Table 2. Estimation of emotion by the association of brain wave and heart rate

Brain wave	Heart rate	Estimated emotion
High arousal	HF predominates	Happy
High arousal	LF predominates	Angry
Low arousal	HF predominates	Sadness
Low arousal	LF predominates	Relax

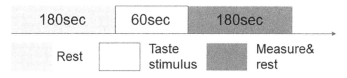

Fig. 3. Experimental paradigm of rest followed by tasting and then measurement of physiological signals by the participants.

All the different taste stimuli were presented sequentially in the order of 1. Control, 2. Umami, 3. Sweet, 4. Salty, 5. Sour and 6. Bitter.

4 Experimental Results

4.1 Analysis of Signals

To analyze the experimental results, we used two different parameters obtained from ECG and EEG. The first one was that obtained using ECG pulse sensor and it was the ratio of LF to HF and this value is a measure of the status of the autonomic nervous system. The second parameter was the one obtained using Neurosky's EEG. The ratio of Attention to Meditation values calculated from the EEG device was used as an index of evaluation. LF is correlated with the sympathetic nerve, thus represents stress. HF correlates with parasympathetic nerves, and it represents relaxation. β waves are associated with attention which in turn represents an awaken status. Meditation correlates with α waves representing relaxation or sleepiness. Table 2 shows the association of brain waves and heart rate and the estimated emotion.

Table 3. The result of questionnaire regarding the taste stimuli; Here the numbers indicate the likeness level with 1 being the most liked and 4 being the most hated tastes.

	Umami	Sweet	Salty	Sour	Bitter
Subject#1	3	3	4	2	4
Subject#2	3	1	3	1	4
Subject#3	1	1	2	2	4

With the above parameters, to estimate emotion, Russel's circumflex model of affection [9] was used. We correlated LF/HF on the valence axis on the model (comfort/discomfort) with Attention/Meditation on the arousal axis. When Attention predominates, high arousal is observed, and when Meditation predominates, low arousal is defined. Similarly, comfort is determined by the heart rate index (LF/HF). When LF was dominant, the dominance of the sympathetic nerve was implied. On the other hand, when HF was dominant, the parasympathetic nervous system played a dominant role.

5   Results

Based on the analyzed information, we found that each subject showed different results. Here, each subject's biological signal is described as follows;

Table 3 shows the experiment result for each subject. The "bitterness" taste could not significantly affect enough biological signals to estimate emotion, therefore, no emotion could be estimated for all subjects. Although individual differences exist in the estimated emotion based on LF/HF and Arousal/Valence ratios, it was found that emotion changed according to the taste. Currently, we are doing experiment to increase the number of subjects and plan to expand to do experiments with a combination of tastes. As from Table 4, except for bitterness which is same for all the subjects, other tastes had different likeness depending on the subject. It would be necessary to study how the likeness would influence the EEG and ECG signals through increasing the number of subjects.

We show attention/mediation (primary Y-axis) and LF/HF values (secondary Y-axis) for all subjects when tasting control, bitter, salty, sour, sweet, and umami in Fig. 4, 5, 6, 7, 8 and 9, respectively. From these figures, it is possible to further analyze the root cause for emotion estimation. For example, Table 3 shows that subject#3's emotion was estimated as "Angry". The emotion is calculated from LF/HF, and attention/meditation.

Table 4. Result summary of attention, meditation, and LF/HF value comparison with baseline value (tasteless water stimulation) for subject 1–3 when stimulated with five basic tastes. Only values with significant effect (**p < .01, *p < .05) are shown. Estimated emotion is shown in the 4th row for each subject data.

Subject		Umami	Sweetness	Saltiness	Sourness	Bitterness
1	Attention		−9.22**	−10.93**		
	Meditation	8.21*	11.75**	10.42**	16.33**	
	LF/HF	1.41**	0.05**	0.22**	4.03**	−0.34**
	Estimated emotion	<u>Sad</u>	<u>Sad</u>	<u>Sad</u>	<u>Sad</u>	
2	Attention	13.15**	12.41**	13.47**	12.45**	6.80*
	Meditation		−14.93**	−9.68**	−24.08**	−9.34**
	LF/HF	2.06**		−0.27*	−0.44**	1.45**
	Estimated emotion	<u>Happy</u>	<u>Happy</u>	<u>Angry</u>	<u>Angry</u>	
3	Attention	8.25**	15.19**		20.10**	11.07**
	Meditation	5.78*				−8.17*
	LF/HF	0.91**		1.39**	3.23**	
	Estimated emotion				<u>Angry</u>	

Taste: Control

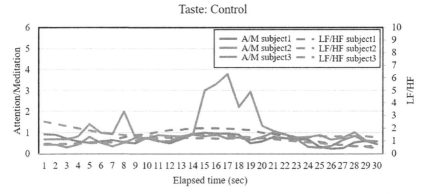

Fig. 4. Attention/Meditation (primary Y-axis) and LF/HF (secondary Y-axis) for all subjects when tasting control stimuli

From these figures, LF/HF value when the subject was tasting sour stimulus is evidently higher than LF/HF value for other tastes. Since LF/HF and meditation value is high, "Angry" is estimated for sour stimulus.

From attention/meditation shown in Fig. 4, 5, 6, 7, 8 and 9, it can be noticed that attention is quite high when subject#3 was tasting control (Fig. 4) and subject#1 was tasting bitter (Fig. 5). Also, the change in LF/HF for sour (Fig. 7), umami (Fig. 9) and bitter (Fig. 5) are large comparing to other taste. It is interesting to note that relax

Taste: Bitter

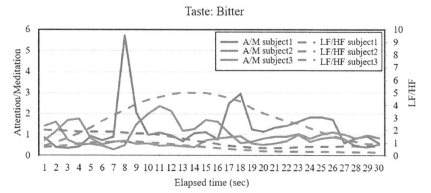

Fig. 5. Attention/Meditation (primary Y-axis) and LF/HF (secondary Y-axis) for all subjects when tasting bitter stimuli.

Taste: Salty

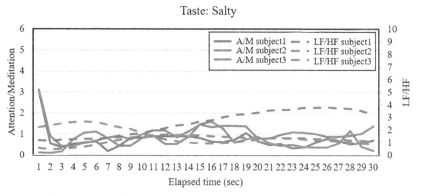

Fig. 6. Attention/Meditation (primary Y-axis) and LF/HF (secondary Y-axis) for all subjects when tasting salty stimuli.

Taste: Sour

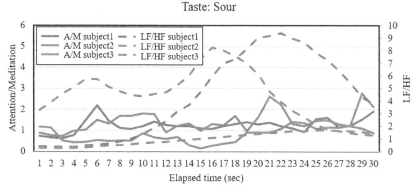

Fig. 7. Attention/Meditation (primary Y-axis) and LF/HF (secondary Y-axis) for all subjects when tasting sour stimuli.

Taste: Sweet

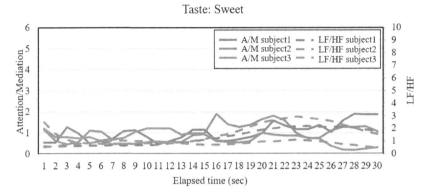

Fig. 8. Attention/Meditation (primary Y-axis) and LF/HF (secondary Y-axis) for all subjects when tasting sweet stimuli.

Taste: Umami

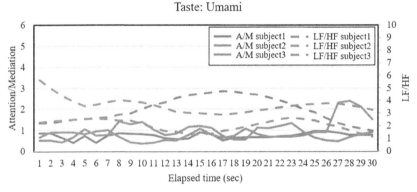

Fig. 9. Attention/Meditation (primary Y-axis) and LF/HF (secondary Y-axis) for all subjects when tasting umami stimuli.

emotion is estimated differently for each subject. Also, the change of A/M and LF/HF for all subjects when tasting sweet (Fig. 8) show some after taste evidence.

6   Conclusion and Future Work

In this paper, we performed a preliminary experiment to prove that human emotion is effected by the taste stimuli. At the same time, the problem of whether the index of LF/HF is suitable for emotion classification has become apparent. In the future, we will examine new indicators of comfort and discomfort, increase the number of experimental collaborators, and group emotional changes according to factors. After that, we will move on to the combined taste sensitivity evaluation.

References

1. Kumakura, H.: Current status, issues and prospects of neuromarketing. Oper. Res. **61**, 421–428 (2016). (in Japanese)

2. Ariely, D., Berns, G.S.: Neuromarketing: the hope and hype of neuroimaging in business. Nat. Rev. Neurosci. **11**, 284–292 (2010)
3. Buchanan, B., Henderson, P.W.: Assessing the bias of preference, detection, and identification measures of discrimination ability in product design. Mark. Sci. **11**, 64–75 (1992)
4. Rangel, A., Camerer, C., Montague, P.R.: A framework for studying the neurobiology of value-based decision making. Nat. Rev. Neurosci. **9**, 545–556 (2008)
5. Small, D.M., Prescott, J.: Odor/taste integration and the perception of flavor. Exp. Brain Res. **166**, 345–357 (2005)
6. Rolls, E.T.: Brain mechanisms underlying flavour and appetite. Philos. Trans. R. Soc. B: Biol. Sci. **361**, 1123–1136 (2006)
7. Sugimoto, K.: Seven relationship between taste stimulation and autonomic nerve activity. Taste (umami) and oral health: aiming for a healthier life. J. Jpn. Assoc. Study Taste Smell **20**, 151–160 (2013). (in Japanese)
8. Kawakami, Y., Komazawa, M., Sugaya, M.: Proposal of emotion classification based on biometric measurement and emotion stabilization method using color (in Japanese). In: 182th HCI Workshop, pp. 1–8 (2019)
9. Russell, J.A.: A circumplex model of affect. J. Personal. Soc. Psychol. **39**, 1161 (1980)

Multimodal Analysis Using Neuroimaging and Eye Movements to Assess Cognitive Workload

Ricardo Palma Fraga[1(✉)], Pratusha Reddy[2], Ziho Kang[1], and Kurtulus Izzetoglu[2]

[1] Department of Industrial and Systems Engineering, University of Oklahoma, Norman, USA
rpalmafr@ou.edu
[2] Drexel University, School of Biomedical Engineering, Science and Health Systems, Philadelphia, USA

Abstract. Air Traffic Control (ATC) specialists work in an environment where the proficient interaction between humans and computer systems is crucial to provide a safe and efficient flow of traffic. The complexity of this system may increase due to planned changes in operator roles and a projected rise in traffic volume. This increase, over an already highly complex system, will exacerbate the mental workload placed on the operator. The emergence of wearable sensors that measure physiological signals enables us to monitor the mental workload in real time without interfering in operational activity. However, the use of a single sensor approach may not provide a comprehensive assessment of cognitive workload while executing a complex task. Therefore, this study implemented a multimodal approach by using two sensors, namely functional near infrared spectroscopy (fNIRS) and eye-tracking, to evaluate the cognitive workload changes experienced by an ATC specialist. Three retired tower controllers with over 20 years of experience, underwent three sessions of experimentation where each individual fulfilled one of the following roles - observer, a Local controller or a Ground controller. During each iteration, the fNIRS and eye tracking sensors were attached to the Local controller while they commanded aircraft through verbal clearances. The task difficulty and complexity were quantified by the number of aircraft and clearances given, respectively. The number of aircraft displayed on the screen increased across time and was positively correlated with oxygenation measures assessed by the fNIRS signals of both the right and left prefrontal cortex. On the other hand, the number of fixations was positively correlated with the number of clearances. These results suggest fNIRS and eye-tracking measures are sensitive to changes in cognitive workload, and indicates that they may be amenable to complement each other for the assessment of the multidimensionality of cognitive workload induced by task difficulty and complexity.

Keywords: Neuroimaging · fNIRS · Eye-tracking · Air Traffic Control · Cognitive workload

© Springer Nature Switzerland AG 2020
D. D. Schmorrow and C. M. Fidopiastis (Eds.): HCII 2020, LNAI 12196, pp. 50–63, 2020.
https://doi.org/10.1007/978-3-030-50353-6_4

1 Introduction

1.1 Background

Air Traffic Control (ATC) specialists are crucial individuals within the system of civil and military aviation in which they maintain safety, while at the same time, expediting the flow of aircraft. In addition to the already established high levels of workplace-related stress (Lesiuk 2008; Finkelman 1994), the environment where they carry out their tasks is expected to become significantly more complex in the near future. This is partially due to numerous long-term projections, estimating that by 2040, future air traffic volume may reach 60 million aircraft annually (FAA 2018) and potentially serve around 10 billion airline passengers (ICAO 2017). In response to the predicted increase in traffic volume, and the continuous efforts by the industry to improve efficiency and safety in air transportation, new technologies and procedures are likely to be adopted in the ATC environment as well. Change to the roles and responsibilities of controllers poses a potential to increase complexity as well. This implies that ATC will impose more mental demands on human operators (Vogt et al. 2006; Ahlstrom and Friedman-Berg 2006) and subject them to a higher risk of human error (Di Nocera et al. 2006).

Due to the potential increase in complexity and workload, there has been a need to conduct a closer, inspection of the interaction between controllers and computers using methods such as NASA-Task Load Index (NASA-TLX) (Hart and Staveland 1988), the Instantaneous Self-Assessment techniques (Kirwan 1997) and the Air Traffic Workload Input Technique (Stein 1985). The aforementioned methods obtain workload ratings post-task, or during interruptions to the task itself. On the other hand, there exists alternative neural and physiological methods that provide direct and real-time evaluation of workload during the task. Additionally, the monitoring of physiological variables has the potential to not only enable a real-time evaluation of an operator's mental state, but to also aid in the informed design of safe and effective adaptive automation systems and future training methods.

Over the last decade, functional Near Infrared Spectroscopy (fNIRS) and eye-tracking are among many devices developed to measure various aspects of cognitive functioning (Bhavsar et al. 2017). fNIRS is an emerging, noninvasive, affordable and portable neuroimaging modality that exploits the optical properties of biological tissues and hemoglobin chromophores in assessing changes in brain activity. It does so by deploying wavelengths between 700 nm to 900 nm, where the chromophores of oxygenated and deoxygenated hemoglobin (HbO and HbR, respectively) are found to be the main absorbers (Jobsis 1977; Delpy et al. 1988). The changes in HbO and HbR are directly associated with changes in brain activity (Villringer and Chance 1997; Izzetoglu et al. 2004), therefore making it an attractive modality to study cognitive performance within the environment of ATC. Alternatively, eye-tracking has established itself as another key methodology that can be used to measure the efficient and timely acquisition of visual information (Bruder and Hasse 2019). The metrics derived from eye movements, such as fixation durations & counts, saccadic lengths, gaze event durations, and pupil dilation have been used to quantify the effort, information processing capabilities, attention, and decision-making ability of participants in numerous complex environments (Otero et al. 2011; Causse et al. 2019; Rudi et al. 2019).

1.2 Multidimensionality of Cognitive Workload

It is a widely accepted fact that one of the major human performance factors in ATC is cognitive workload (Edwards 2013; Ball et al. 2007), which is affected by the airspace design and traffic requirements (Durso and Manning 2008). The number of aircraft under control has been studied numerous times and shown to be associated with mental workload measures (Ahlstrom and Friedman-Berg 2006). However, it is not the only environmental variable capable of increasing cognitive workload and may not necessarily represent the only quantifiable measure of complexity for the ATC task (Kirwan et al. 2001; Athenes et al. 2002). There exist additional operator-dependent factors, such as strategy, that play an important role in quantifying cognitive workload. How a controller chooses to prioritize tasks, or compensatory strategies used to respond to workload fluctuation, all influence an ATC specialist's cognitive workload (Koros et al. 2003).

This multidimensional aspect of cognitive workload can be defined by the interaction between three categories: (1) drivers (activity-based estimators), which evaluate the prescription and quantity of information to process; (2) mediators (operator-based estimators) where changes of strategy are quantified; (3) indicators (activity or operator-based estimators) which include performance metrics, the participant's subjective experience with the task, and physiological measures (Kostenko et al. 2016). In the ATC setting, the drivers are represented by the task difficulty (e.g. number of aircraft or weather conditions), mediators portray how controllers strategize with the complexity of the task (e.g. number of commands), and indicators can be characterized by quantifiable performance measures (e.g. neural and psychophysiological measures).

fNIRS and eye-tracking have been widely used to measure controller workload (e.g. Truschzinski et al. 2018; Marchitto et al. 2016; Harrison et al. 2014; Tsai et al. 2012). Results from the fNIRS studies indicated that oxygenation measures were sensitive to task load changes and correlated highly with behavioral performance measures (Izzetoglu et al. 2004 and Izzetoglu and Richards 2019; Ayaz et al. 2012; Reddy et al. 2018). Similarly, eye-tracking metrics such as pupil dilation, fixation number & duration, and saccades have shown to be closely related with cognitive workload (van der Wel and van Steenbergen 2018; Eckstein et al. 2017).

1.3 Multimodal Approach

The utility & reliability of a multimodal approach is imperative to achieve well-rounded and concise results, as one singular approach may fail to capture all the critical elements of cognitive workload. For example, researchers have shown that application of a multimodal methodology has helped in understanding how levels of automation impact cognitive workload (Evans and Fendley 2017), assess multiple factors that influence situational awareness (Friedrich et al. 2018), and identify differences between experts and novices (İşbilir et al. 2019). We expand on this topic by investigating the correlations that exist between the variables captured from a multimodal approach, consisting of fNIRS and eye-tracking, alongside behavioral measures. These have been defined by considering cognitive workload as a multidimensional construct and quantified in terms of the number of aircraft (task difficulty) & number of commands (task complexity).

2   Methodology

2.1   Participants

Three retired ATC specialists, each with over 20 years of experience as tower controllers, participated in the experiment by switching and performing three different
roles: (1) observer; (2) operating the Local control position; (3) operating the Ground
control position. The participants were recruited with the help of the Federal Aviation Administration's Civil Aerospace Medical Institute (CAMI) in Oklahoma City,
Oklahoma.

2.2   Simulator and Sensors Used

The simulated air and ground traffic experienced at an airport was displayed via eight
55" HD (1080p) screens. Three separate monitors were used for the Airport Surface
Detection Equipment (ASDE), Bright Radar Indicator Tower Equipment (BRITE) and
Status Information Area (Fig. 1). The simulator software – MaxSim, developed by Adacel
Systems Inc – was used to generate the scenario. Participants instructed the aircraft
through verbal commands called clearances, which were processed by voice recognition
software embedded in the simulator. The recognition software responded accordingly
with a simulated voice reply and aircraft behavior consistent with the command. The
simulator had a pseudo-pilot, whose role was to monitor the interpretations of the voice
commands given by controllers and take the necessary steps to correct any misrecognized
or unrecognized utterances.

   For data collection purposes, a pair of Tobii Pro Glasses II (100 Hz) (REF for
company) was used to capture the eye movements, while fNIRS 1000 (2 Hz, 10 detectors,
4 LEDs that operated at 730 nm, 850 nm, and ambient wavelengths) created by fNIRS
Systems, LLC (REF for company) was used to measure raw light intensity data from
the prefrontal cortex (PFC).

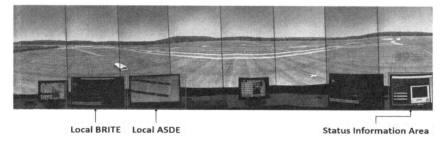

Local BRITE   Local ASDE   Status Information Area

Fig. 1. The simulation environment was presented to the participants during the experiment. The
additional displays that can be used by the local controller are highlighted in red. (Color figure
online)

2.3 Scenario and Task

All participants underwent a single 32-min scenario developed by the Federal Aviation Administration (FAA). The scenario consisted of 33 aircraft, with 19 arrivals and 14 departures. Within this scenario, participants carried out duties (issue landing and takeoff instructions to pilots, monitor and direct the movement of aircraft on the ground and in the air) of a Local controller. No pre-determined conflicts were incorporated into the scenario and the weather conditions (clear blue skies) remained consistent throughout the scenario. This procedure was chosen in order to elicit normative behavior of controlling traffic in a tower control environment from the participants.

2.4 Data Analysis

The eye-tracking data was subjected to the automated I-VT filter (Komogortsev and Karpov 2013) within Tobii to extract the fixations and saccades. Here, fixations were defined at a threshold of at least 60 ms, while everything else was considered a saccade. The data obtained from these eye-movements included: left and right pupil dilation, number of fixations, gaze event duration, and the sequential saccadic distance. To factor individual differences in natural pupil dilation, those estimates were normalized by calculating a grand pupil dilation mean across all participants, which was then used to divide each pupil dilation data point (Engelhardt et al. 2010).

On the other hand, the fNIRS signal was first corrected for light leaks by simply subtracting ambient signal from the other two signals (730 nm, and 850 nm). The signals were then low pass filtered to remove instrument noise, physiological noise and motion artifacts (Izzetoglu et al. 2005; Reddy et al. 2018). Then, modified Beer-Lambert Law was used to calculate the oxygenated (HbO) and deoxygenated-hemoglobin (HbR) changes at each channel (Villringer and Chance 1997). Using HbR and HbO measures, oxygenation (Oxygenation $=$ HbO $-$ HbR) and total hemoglobin (Hbtot $=$ HbO $+$ HbR) were derived. Lastly, samples that were three standard deviations above the expected values were classified as outliers and removed (0.34%) from further analysis.

Task difficulty and complexity factors were quantified using number of aircraft and number of clearances measures. The former was identified by summing the number of aircraft managed by the Local controller as seen on the ASDE and BRITE radars. This approach is different than simply considering the total number of aircraft present in the scenario at any particular instant, as not all of them fall under the direct command of the Local controller (e.g. aircraft taxiing from their respective gate to the runways are managed by the Ground controller, but they also appear on the radar displays). The latter was determined by counting clearances given per time bin associated with each aircraft condition. Additional information regarding the type of clearance given, and the aircraft that the clearance was given to were also noted.

The times at which the task difficulty changed were used to bin eye-tracking and fNIRS measures. The resulting average measures of the extracted features were assessed for correlation between each other. One-way ANOVA or non-parametric Kruskal–Wallis test was used to determine the difference between neural and psychophysiological metrics across the task load measures. Post-hoc Tukey or Mann Whitney tests were used to determine where the differences come from. An alpha level of 0.05 was used as the significance criteria for tests conducted.

Data from the first three minutes were removed from the analysis, because participants used this time to become familiar with the task environment. In addition, the data segment associated with 12-aircraft was removed as only one participant controlled 12 aircraft simultaneously.

3 Results

3.1 Correlation Between Task, Behavioral and Physiological Measures

A correlation analysis was conducted to investigate the relationship between task factors, neural and psychophysiological metrics and the results of this analysis are seen in (Fig. 2). Average oxygenation measures from both left (Channels: 3, 4, 5, 6) and right (Channels: 11, 12, 13, 14) prefrontal cortex (PFC) negatively correlated with normalized left & right pupil dilation ($r = -0.47$ & -0.48, with both having a p-val < 0.01), and gaze duration ($r = -0.38$, p-val < 0.01), but not with other eye-tracking measures. Both left and right hemisphere cortical oxygenation increased as the number of aircraft increased. However, no trend was observed between the oxygenation changes and the number of clearances. Alternatively, the number of fixations increased with the number of clearances, but not with the number of aircraft. Besides the fixation number, no other eye-tracking metrics revealed any correlation with the task factors.

3.2 Interaction Between Task and Behavioral Measures

The correlation between the number of aircraft and number of clearnaces was low ($r = 0.36$), when it was expected to be high. To understand the relationship between these two behavioral metrics better, we investigated the temporal profiles of each variable. The results of this analysis are shown in Fig. 3, where the number of aircraft managed by the controller changed quadratically with time, while the number of clearances changed cubically. By overlaying the two plots or plotting number of aircraft vs number of clearances, three phases of task demand can be identified. The first phase (3 to 9 min) shows an increase in both number of clearances and number of aircraft. The second phase (9 to 21 min) is defined by an increase in the number of aircraft, but no change in number of clearances. Finally, the last phase (21 to 33 min) shows an increase in the number of clearances, while the number of aircraft does not change. These three phases were used to define three levels of task demand - Low, Medium and High. A one-way ANOVA was used to assess whether the number of clearances differed across groups or not. Results indicated significant difference across task demand groups ($F_{2,64} = 3.329$, $p = 0.04$), with significant difference between low and high group ($p = 0.01$).

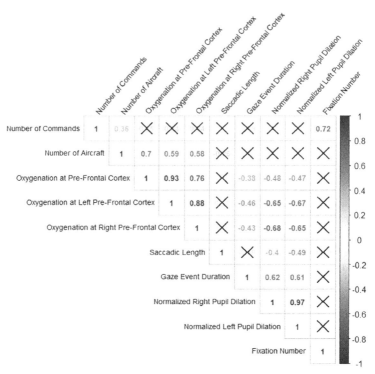

Fig. 2. Correlation plot (R package: corrplot, Wei and Simko 2017) between the behavioral measures, fNIRS and eye-tracking metrics. The X's represent non-significant correlations (p-val > 0.01).

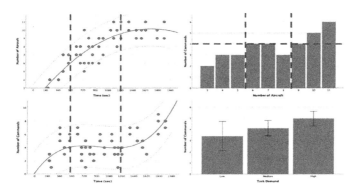

Fig. 3. Change in number of aircraft across time (left, top) and change in number of clearances across time (left, bottom). Median changes in the number of clearancess relative to the number of aircraft (left, top). Average number of clearances per low, medium and high load conditions. Blue lines represent the partitions defined by multidimensional workload model. Error bars are ±2 standard error of the mean (SEM). (Color figure online)

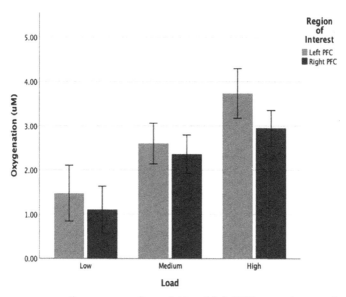

Fig. 4. Average oxygenation measures from right and left PFC across low, medium and high multidimensional load conditions. Error bars are ±2 standard error of the mean (SEM).

3.3 Effects of Changes in Drivers and Mediators on Neural and Psychophysiological Measures

After checking for normality and parametric assumptions, oxygenation measures from the prefrontal cortex (PFC) were submitted to a one-way ANOVA to assess the effect of the number of aircraft. Significant effects of the aircraft number on oxygenation measures were reported ($F_{8,56} = 6.95$, p < 0.01). Post hoc comparisons using Tukey test indicated no differences between aircraft conditions ranging from 3 to 8 and 9 to 11. However, conditions 9 to 11 revealed significantly higher oxygenation measures in comparison to 3 and 4, while condition 11 was higher than 6 and 7. This result suggested three levels of task load, which were defined as low (3 to 5 aircrafts), medium (6 to 8 aircrafts), and high (9 to 11 aircrafts). After establishing these three groups, we conducted a one-way ANOVA again to assess the effect of multidimensional task load on oxygenation from left and right PFC. The results as shown in Fig. 4 indicate significant effect of workload on the left ($F_{2,62} = 10.61$, p = 0.00) and on the right ($F_{2,62} = 10.29$, p = 0.00) PFC, with differences across all load levels per region of interest.

A non-parametric Kruskal–Wallis test was used to examine the effect of workload on eye-tracking measures (fixation number, gaze event duration and saccadic length). This was performed since the measures failed to meet normality and parametric assumptions. Only saccadic length showed significant decrease ($\chi_2 = 9.3$, p = 0.01, df = 2) across task load. A post-hoc Mann Whitney test suggested that these decreases were significant between high and other load groups but not low and medium groups. An increasing trend was seen in fixation number and decreasing trend was observed in the gaze event duration (Fig. 5).

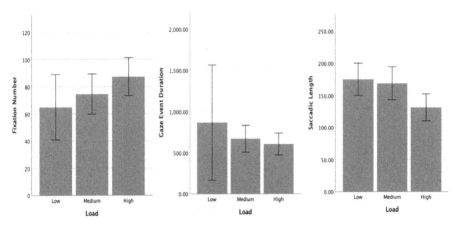

Fig. 5. Average eye-tracking measures (fixation number, gaze event duration and saccadic length) across low, medium and high multidimensional load conditions. Error bars are ±2 SEM.

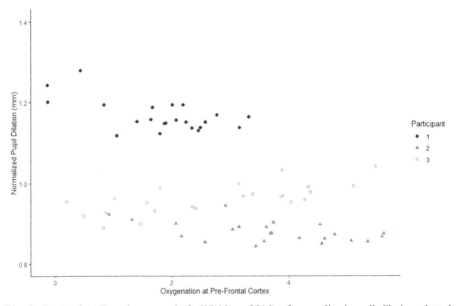

Fig. 6. Scatterplot (R package: ggplot2, Wickham 2016) of normalized pupil dilation plotted versus oxygenation levels at the PFC.

3.4   Changes in Neural and Physiological Measures Relative to Each Other

The correlation coefficient of $r = 0.97$ between normalized left and right pupil dilation points towards a common causal factor (Hansen et al. 2018). As such, normalized right and left pupil dilation were averaged and will be referred to as the normalized pupil dilation. The oxygenation changes from the PFC were plotted against the normalized pupil dilation across participants (Fig. 6) to observe the different workload capabilities elicited through the scenario. Here, participant 1 and 2 had a relatively decreasing trend, but the former maintained the lowest levels of oxygenation and the highest normalized pupil dilation, while in the latter, the opposite trend was observed. On the other hand, participant 3 had an increasing trend but maintained similar characteristics as participant 2. Additionally, the normalized pupil dilation levels appear to remain relatively constant without strong changes.

4 Discussion

Aircraft density and air traffic complexity have long been known to be two major factors involved in workload imposed on ATC specialists (Kirwan et al. 2001; Athenes et al. 2002). This study aimed to quantify these factors' influence on workload using a multimodal approach.

The results indicate that (i) eye-tracking and fNIRS measures are sensitive to increases in either task complexity or difficulty, but not to both; (ii) task difficulty and complexity are non-linearly related; (iii) increase in task demands, quantified by the interaction between task difficulty and complexity, cause an increase in both fNIRS and eye-tracking measures (iv) implementation of behavioral and physiological measures could serve as metric for assessing expertise even within a skilled ATC officer group.

These preliminary results revealed that different modalities could be sensitive to different task demand factors. In this study, eye-tracking measures (fixation number) increased linearly with task complexity (number of clearances), while fNIRS measures (oxygenation) across both hemispheres increased linearly with task difficulty (number of aircraft). These findings are in agreement with unimodal fNIRS or eye-tracking studies that assessed workload of controllers (Ayaz et al. 2012; Izzetoglu and Richards 2019; Reddy et al. 2018; van der Wel and van Steenbergen 2018; Tsai et al. 2012). Specifically, fNIRS studies assessing cognitive workload in ATC specialists, used the number of aircraft as an indicator of task demand and reported that as the demand increased, so did blood oxygenation response in both right and left hemisphere. Eye-tracking studies have shown that there exists a positive correlation between fixation number and workload (Ha et al. 2006). In this case, similar results were found as there exists a strong correlation between the number of fixations and the number of clearances ($r = 0.72$). The finding as to why fNIRS measures correlated with number of aircrafts, but not with number of clearances, and vice-versa for eye-tracking measures, calls for further study with more participants. Considering the limited sample size, this is still a critical finding and should be investigated further.

Task complexity, assumed in this study by the number of clearances, did not behave linearly with the task difficulty as defined by number of aircraft. At the beginning of the scenario there was an increase in aircraft, which was correlated with an increase in clearances. This positive correlation may be due to the fact that an experienced ATC officer can divide his/her attention to execute multiple simultaneous tasks. Following this trend, there was no change in clearances even though the number of aircraft increased, which may be explained by the known fact that attendance to multiple tasks is limited in number by the human perceptual system (Wickens and Hollands 2003). In order to prevent an overload, an experienced controller would take note of information (working memory) (Wickens 2002), but only attend to a set number of aircraft at a given moment.

The final finding of this study is related to expertise. We discovered that the first participant was much more tolerable to increases in task demand in comparison to the other two participants. Specifically, he had no significant differences in behavioral performance when compared with other participants but showed significantly lower increase in fNIRS and higher increase in pupil dilation. This result could be in line with the hypothesis that expertise tends to be associated lower brain activity relative to novices, particularly in PFC areas (Milton et al. 2004). The reasoning as to why this happens can be explained by the fact that PFC is involved in higher order functioning, so once one becomes an expert her/his PFC frees up space for novel incoming stimuli or task demand. However, this evaluation of expertise is dependent on the expectation that behavioral performance positively correlated with task demand, as decrease in performance and brain activity would be indicative of a participant not engaging with the task or being overloaded (Izzetoglu et al. 2004).

This study was conducted under a number of limitations. This is the first time, as far as the authors are aware, that the number of clearances was used as a measure of complexity that was compared with physiological measures, therefore the use of this metric as a task demand factor needs to be investigated and thus validated with a larger sample size. The resulting distinction between participant 1 from the other two in an expertise context may be insignificant if the experiment were to be repeated more than once. Lastly results presented in this study need to be supported by enrolling a larger sample population, considering the fact that other multi-modal studies have reported contradictory results. For instance, Hogervorst et al. 2014 investigated which physiological variables provide the most accurate information on cognitive workload and found non-significant trends of improved accuracy for a combination of electroencephalogram (EEG) and eye-tracking over EEG alone, 91% and 86% accuracy respectively. Another study identified that neurophysiological measures are more sensitive, in comparison to eye-tracking metrics, when discriminating the impact that task difficulty and complexity have in a high-fidelity driving scenario (Flumeri et al. 2018). In contrast, Bernhardt et al. (2019), reached the conclusion that EEG workload and pupil dilation are not correlated to each other, a finding also shared by Matthews et al. (2015), and hypothesized that it may be that pupil dilation may not be a direct measure of cognitive workload. Regardless, fNIRS and eye-tracking are portable, safe, affordable and unobtrusive systems, which can present a potential multi-sensor approach to a multidimensional evaluation of an ATC specialist's cognitive workload.

Acknowledgment. The authors would like to gratefully thank and recognize Dr. Jerry Crutchfield (FAA – Civil Aeronautical Medical Insitute. Oklahoma City, Oklahoma) for his continuous support through the experimental protocol, data acquisition phases of the experiment; and for facilitating access to the simulation environment. In addition, his comments throughout the writing process of this paper were invaluable.

References

Lesiuk, T.: The effect of preferred music listening on stress levels of air traffic controllers. Arts Psychother. **35**, 1–10 (2008)

Finkelman, J.M.: A large database study of the factors associated with work-induced fatigue. Hum. Factors **36**, 232–243 (1994)

Federal Aviation Administration: Forecasts of IFR aircraft handled by FAA air route traffic control centers FY 2017-2040 (2018)

ICAO: The World of Air Transport in 2017. https://www.icao.int/annual-report-2017/Pages/the-world-of-air-transport-in-2017.aspx. Accessed 09 Feb 2020

Vogt, J., Hagemann, T., Kastner, M.: The impact of workload on heart rate and blood pressure in en-route and tower air traffic control. J. Psychophysiol. **20**, 297–314 (2006)

Ahlstrom, U., Friedman-Berg, F.J.: Using eye movement activity as a correlate of cognitive workload. Int. J. Ind. Ergon. **36**(7), 623–636 (2006)

Di Nocera, F., Fabrizi, R., Terenzi, M., Ferlazzo, F.: Procedural errors in air traffic control: effects of traffic density, expertise, and automation. Aviat. Space Environ. Med. **77**, 639–643 (2006)

Hilburn, B., Jorna, P.G., Byrne, E.A., Parasuraman, R.: The effect of adaptive air traffic control (ATC) decision aiding on controller mental workload. In: Mouloua, M., Koonce, J. (eds.) Human-Automation Interaction: Research and Practice, pp. 84–91. Erlbaum, Mahwah (1997)

Hart, S.G., Staveland, L.E.: Development of NASA-TLX (Task Load Index): results of empirical and theoretical research. In: Hancock, P.A., Meshkati, N. (eds.) Advances in Psychology, Human Mental Workload, pp. 139–183. Amsterdam, Elseiver (1988)

Langan-Fox, J., Sankey, M.J., Canty, J.M.: Human factors measurement for future air traffic control systems. Hum. Factors J. Hum. Factors Ergon. Soc. **51**(5), 595–637 (2009)

Bhavsar, P., Srinivasan, B., Srinivasan, R.: Quantifying situation awareness of control room operators using eye-gaze behavior. Comput. Chem. Eng. **106**, 191–201 (2017)

Bruder, C., Hasse, C.: Differences between experts and novices in the monitoring of automated systems. Int. J. Ind. Ergon. **72**, 1–11 (2019)

Otero, S.C., Weekes, B.S., Hutton, S.B.: Pupil size changes during recognition memory. Psychophysiology **48**(10), 1346–1353 (2011)

Causse, M., Lancelot, F., Maillant, J., Behrend, J., Cousy, M., Schneider, N.: Encoding decisions and expertise in the operator's eyes: using eye-tracking as input for system adaptation. Int. J. Hum Comput Stud. **125**, 55–65 (2019)

Rudi, D., Kiefer, P., Raubal, M.: The instructor assistant system (iASSYST) utilizing eye tracking for commercial aviation training purposes. Ergonomics **63**(1), 61–78 (2019)

Edwards T.: Human performance in air traffic control. Dissertation, University of Nottingham (2013)

Ball, M., Barnhart, C., Nemhauser, G., Odoni, A.: Air transportation: irregular operations and control. In: Barnhart, C., Laporte, G. (eds.) Handbooks in Operations Research and Management Science, vol. 14, pp. 1–67. Elsevier, Amsterdam (2007)

Durso, F.T., Manning, C.A.: Air traffic control. Rev. Hum. Factors. Ergon. **4**, 195–244 (2008)

Kostenko, A., Rauffet, P., Coppin, G.: A dynamic closed-looped and multidimensional model for mental workload evaluation. IFAC - Papers On Line **49**(19), 549–554 (2016)

Ayaz, H., Shewokis, P.A., Bunce, S., Izzetoglu, K., Willems, B., Onaral, B.: Optical brain monitoring for operator training and mental workload assessment. NeuroImage **59**(1), 36–47 (2012)

Evans, D.C., Fendley, M.: A multi-measure approach for connecting cognitive workload and automation. Int. J. Hum Comput Stud. **97**, 182–189 (2017)

Friedrich, M., Biermann, M., Gontar, P., Biella, M., Bengler, K.: The influence of task load on situation awareness and control strategy in the ATC tower environment. Cogn. Technol. Work **20**(2), 205–217 (2018)

İşbilir, E., Çakır, M.P., Acartürk, C., Tekerek, A.Ş.: Towards a multimodal model of cognitive workload through synchronous optical brain imaging and eye tracking measures. Front. Hum. Neurosci. **13**, 1–13 (2019)

Truschzinski, M., Betella, A., Brunnett, G., Verschure, P.F.M.J.: Emotional and cognitive influences in air traffic controller tasks: an investigation using a virtual environment. Appl. Ergon. **69**, 1–9 (2018)

Marchitto, M., Benedetto, S., Baccino, T., Cañas, J.J.: Air traffic control: ocular metrics reflect cognitive complexity. Int. J. Ind. Ergon. **54**, 120–130 (2016)

Harrison, J., et al.: Cognitive workload and learning assessment during the implementation of a next-generation air traffic control technology using functional near-infrared spectroscopy. IEEE Trans. Hum.-Mach. Syst. **44**(4), 429–440 (2014)

Tsai, M.-J., Hou, H.-T., Lai, M.-L., Liu, W.-Y., Yang, F.-Y.: Visual attention for solving multiple-choice science problem: an eye-tracking analysis. Comput. Educ. **58**(1), 375–385 (2012)

van der Wel, P., van Steenbergen, H.: Pupil dilation as an index of effort in cognitive control tasks: a review. Psychon. Bull. Rev. **25**(6), 2005–2015 (2018)

ICAO.: Doc 9328: Manual of Runway Visual Range Observing and Reporting Practices. International Civil Aviation Organization, 3rd edn. (2005)

van Schaik, F.J., Roessingh, J.J.M., Lindqvist, G., Fält, K.: Assessment of visual cues by tower controllers, with implications for a remote tower control centre. IFAC Proc. Vol. **43**(13), 123–128 (2010)

Ha, C.H., Kim, J.H., Lee, S.J., Seong, P.H.: Investigation on relationship between information flow rate and mental workload of accident diagnosis tasks in NPPs. IEEE Trans. Nucl. Sci. **53**(3), 1450–1459 (2006)

Di Flumeri, G., et al.: Brain–computer interface-based adaptive automation to prevent out-of-the-loop phenomenon in air traffic controllers dealing with highly automated systems. Front. Hum. Neurosci. **13**, 296 (2019)

Bernhardt, K.A., et al.: The effects of dynamic workload and experience on commercially available EEG cognitive state metrics in a high-fidelity air traffic control environment. Appl. Ergon. **77**, 83–91 (2019)

Matthews, G., Reinerman-Jones, L.E., Barber, D.J., Abich, J.: The psychometrics of mental workload: multiple measures are sensitive but divergent. Hum. Factors **57**(1), 125–143 (2015)

Hogervorst, M.A., Brouwer, A.M., van Erp, J.B.F.: Combining and comparing EEG, peripheral physiology and eye-related measures for the assessment of mental workload. Front. Neurosci. **8**, 1–15 (2014)

Komogortsev, O.V., Karpov, A.: Automated classification and scoring of smooth pursuit eye movements in the presence of fixations and saccades. Behav. Res. Methods **45**, 203–215 (2013)

Wei, T., Simko, V.: R package "corrplot": visualization of a correlation matrix (version 0.84) (2017)

Engelhardt, P.E., Ferreira, F., Patsenko, E.G.: Rapid communication pupillometry reveals processing load during spoken language comprehension. Q. J. Exp. Psychol. **63**(4), 639–645 (2010)

Wickham, H.: R Package "ggplot2": Elegant Graphics for Data Analysis. Springer, New York (2016). https://doi.org/10.1107/978-0-387-98141-3

Hansen, J.P., Hardenberg, D., Biermann, F., Bækgaard, P.: A gaze interactive assembly instruction with pupillometric recording. Behav. Res. Methods **50**(4), 1723–1733 (2018)

Eckstein, M.K., Guerra-Carrillo, B., Miller Singley, A.T., Bunge, S.A.: Beyond eye gaze: what else can eyetracking reveal about cognition and cognitive development? Dev. Cogn. Neurosci. **25**, 69–91 (2017)

Alemdag, E., Cagiltay, K.: A systematic review of eye tracking research on multimedia learning. Comput. Educ. **125**, 413–428 (2018)

Athènes, S., Averty, P., Puechmorel, S., Delahaye, D., Collet, C.: ATC complexity and controller workload: trying to bridge the gap. www.aaai.org. Accessed 10 Feb 2020

Delpy, D.T., Cope, M., van der Zee, P.: Estimation of optical path length through tissue from direct time of flight measurement. Phys. Med. Biol. **33**(12), 1433–1442 (1988)

Herrmann, M.J., et al.: D4 receptor gene variation modulates activation of prefrontal cortex during working memory. Eur. J. Neurosci. **26**(10), 2713–2718 (2007)

Hilburn, B., Jorna, P.G.A.M.: Workload and air traffic control. In: Human Factors in Transportation. Stress, Workload, and Fatigue, pp. 384–394 (2001)

Izzetoglu, M., Izzetoglu, K., Bunce, S., Ayaz, H., Devaraj, A., Onaral, B.: Functional Near-Infrared Neuroimaging. IEEE Trans. Neural Syst. Rehabil. Eng. **13**(2), 153–159 (2005)

Jobsis, F.F.: Noninvasive, infrared monitoring of cerebral and myocardial oxygen sufficiency and circulatory parameters. Science **198**(4323), 1264–1267 (1977)

Kirwan, B.: Validation of human reliability assessment of techniques: part 1 - Validation issues. Saf. Sci. **27**(1), 25–41 (1997)

Kirwan, B., Scaife, R., Kennedy, R.: Investigating complexity factors in UK air traffic management. In: Engineering Psychology and Cognitive Ergonomics, pp. 189–195 (2017)

Stein, E.S.: Air traffic controller workload: an examination of workload probe. Atlantic City International Airport: Federal Aviation Administration Technical Center. (DOT/FAA/CT-TN84/24) (1985)

Koros, A., Rocco, P.S., Panjwani, G., Ingurgio, V., D'Arcy, J.F.: Complexity in air traffic control towers: a field study. Part 1: complexity factors (2003)

Martijn Jansma, J., Ramsey, N.F., Coppola, R., Kahn, R.S.: Specific versus nonspecific brain activity in a parametric N-back task. Neuroimage **12**, 688–697 (2000)

Matsuo, K., et al.: Prefrontal hyperactivation during working memory task in untreated individuals with major depressive disorder. Mol. Psychiatry **12**(2), 158–166 (2007)

Milton, J.G., Small, S.S., Solodkin, A.: On the road to automatic: dynamic aspects in the development of expertise. J. Clin. Neurophysiol. **21**(3), 134–143 (2004)

Veltman, D.J., Rombouts, S.A.R.B., Dolan, R.J.: Maintenance versus manipulation in verbal working memory revisited: an fMRI study. NeuroImage **18**(2), 247–256 (2003)

Villringer, A., Chance, B.: Non invasive optical spectroscopy and imaging of human brain function. Trends Neurosci. **20**(10), 435–442 (1997)

Wickens, C.D., Hollands, J.G.: Engineering Psychology and Human Performance, 3rd edn. Prentice Hall, Upper Saddle River (2000). N. Roberts & B. Webber

Wickens, C.D.: Situation awareness and workload in aviation. Curr. Dir. Psychol. Sci. **11**(4), 128–133 (2002)

Izzetoglu, K., Bunce, S., Onaral, B., Pourrezaei, K., Chance, B.: Functional optical brain imaging using near-infrared during cognitive tasks. Int. J. Hum.-Comput. Interact. **17**(2), 211–227 (2004)

Reddy, P., Richards, D., Izzetoglu, K.: Cognitive performance assessment of UAS sensor operators via neurophysiological measures. Front. Hum. Neurosci. **12** (2018)

Izzetoglu, K., Richards, D.: Human performance assessment: evaluation of wearable sensors for monitoring brain activity. In: Vidulich, M., Tsang, P. (eds.) Improving Aviation Performance through Applying Engineering Psychology: Advances in Aviation Psychology, 1st edn, pp. 163–180. CRC Press, Boca Raton (2019)

Challenges and Prospects of Emotional State Diagnosis in Command and Control Environments

Alina Schmitz-Hübsch$^{(\boxtimes)}$ ⓘ and Sven Fuchs ⓘ

Fraunhofer Institute for Communication, Information Processing and Ergonomics FKIE,
53343 Wachtberg, Germany
{alina.schmitz-huebsch,sven.fuchs}@fkie.fraunhofer.de

Abstract. The present investigation examines the correlations of emotional user states and performance in command and control environments. The aim is to gain insights into whether and how the integration of the emotional state into a diagnostic component of an adaptive human-machine system can take place. While positive valence is expected to be associated with high performance (Hypothesis 1), negative valence is assumed to be associated with low performance (Hypothesis 2). In a laboratory experiment, a command and control task in the domain of anti-air warfare was performed. Emotional valence was assessed with a software for recognition of emotional face expressions (Emotient FACET). To measure performance, we assessed performance decrements that occurred whenever a subtask was not accomplished in time. Data from 22 participants were used for the analysis (45% female, 17–52 years, M = 30.96, SD = 9.76). Regression analyses were conducted to test the hypotheses. Contrary to expectations, there were no correlations between emotional valence and performance at the group level. Even though empirical results failed to support the hypotheses, significant correlations between positive valence and performance were found for 36% of the subjects, 45% for negative valence and performance. These results indicate that individual models are necessary for the analysis of emotional user state. We discuss practical implications and suggest improvements to the paradigm.

Keywords: Affect-adaptive systems · Affective user state · Adaptive automation · Command and control

1 Introduction

Increasing automation in modern work environments does not necessarily improve work performance. With high levels of automation, new problems can arise. Loss of cognitive or motor skills, inadequate trust in the automation, or loss of situation awareness can decrease performance [1, 2]. To address these issues, adaptively automated systems are being developed that adapt technical functions or interactions to conditions such as the situational context or the current workload. Furthermore, the system may consider the current performance of the user, his state, or well-being in order to adapt accordingly.

© Springer Nature Switzerland AG 2020
D. D. Schmorrow and C. M. Fidopiastis (Eds.): HCII 2020, LNAI 12196, pp. 64–75, 2020.
https://doi.org/10.1007/978-3-030-50353-6_5

Some user states like critical fatigue [3] or incorrect focus of attention [4] can impair performance. Fuchs et al. [5] developed a prototype adaptive system that considers the situational context, the user's performance, and different mental user states. Based on this diagnosis, context-adequate forms of support, so called adaptations or adaptation strategies can be triggered.

The work reported in this paper aims at investigating the correlation between emotional user states and performance in the context of *Command and Control (C2)*. Results will be used to gain insight about a possible integration of the emotional state in Fuchs et al.'s adaptive system. If a correlation between emotional state and performance can be found, it could be useful for the adaptive system to consider emotional states. For example, if negative emotional states prove to be associated with performance decrements, a future adaptive system could aim at mitigating these negative emotions to improve performance.

First, the relationship between emotional user state and performance will be examined. In literature, positive emotional states are associated with high performance in human machine interaction [6–8]. Negative emotional states have been found to correlate with low performance [9–11]. To examine these correlations in C2 environments, an experiment using an anti-air warfare task is reported. A software for emotion recognition (Emotient FACET [12]) detected emotional valence in facial expressions. As a performance metric we used the number of incomplete tasks whenever the diagnostic component reported performance decrements (which occurred when a subtask was not completed within a predefined time limit).

1.1 Background

Command and control is a military term that describes the knowledgeable exercise of authority in accomplishing military objectives and goals [13]. The interaction of human, physical, and information resources allows problem solving using organizational and technical attributes and processes [14]. Especially in safety-critical domains, performance decrements can have serious consequences and need to be avoided at all times [15]. In the present investigation, correlations of performance and emotional user states are investigated. In a future adaptive system, the insights gained might be used to mitigate performance decrements related to emotional states.

Former investigations of human-machine systems in different areas show undesirable consequences of negative emotions. Kontogiannis [9] found that operators who were stressed, upset, or frustrated did not achieve their optimal level of performance in a complex task environment. Madhavan et al. [10] showed that taking over highly automatized systems and switching to the manual mode was more difficult in fear inducing situations. Szameitat et al. [11] examined the consequences of delays and interruptions on the emotional state and performance. In the delay condition, a higher error rate, longer reaction time, and negative emotional valence were found.

Negative emotions may impair performance in various ways [16]:

1. Negative emotions distract attention and decrease processing efficiency.
2. They can change the way information is processed, for example by increasing fault tolerance.

3. At the extreme, goal-driven behavior can be interrupted because operators doubt meaningfulness and feasibility of their mission.
4. State regulation may be disturbed which facilitates over-reactivity and may reduce task performance by a high adrenalin level or muscle stiffness.
5. Psychosomatic disorders like headaches or stomach aches may occur that impair attention or motivation.
6. Negative emotions may exacerbate recovery and cause insomnia that reduce performance long term.

Only few studies have examined the correlation of emotion and performance in C2 environments. In her master's thesis, Panganiban [17] used an air battle management task and found better performance in the defense of air contacts when operators scored high on trait-anxiety. However, this effect was not found for the emotional state anxiety.

It is likely that findings from the driving domain can be transferred to C2, as similar requirements exist in both areas. First, both are safety-critical. Therefore, attentiveness, avoidance of errors, and high task performance are crucial. Secondly, as task automation increases in both domains, tasks require observation rather than execution. Eyben et al. [6] found that negative emotions with a low level of arousal can have negative consequences on driving performance. Frustration and sadness, for example, impaired attention and were accompanied by resignation and passivity and thereby increased reaction time. Stecklov and Goldstein [18] examined the influence of terrorist attacks on driving performance in Israel. They showed an increase of traffic accidents by 35% the third day after the attack. Zimasa, Jamson and Henson [19] induced sadness, happiness, and neutral emotions. They discovered that sad study participants had the highest fixation durations and longest reaction times in hazard recognition in traffic. Negative emotions with a high level of arousal like anger compromise driving performance and increase the risk of causing an accident [20]. Even low levels of aggression can distract the driver's attention, impair concentration, and therefore increase the risk of an accident [21]. Literature clearly answers the question for the optimal emotional user state in driving with the statement "Happy drivers are better drivers" [6, p. 2]. Jones and Jonsson [7] and Groeger [8] also suspect that positive valence contributes to better driving performance. Data of current studies is ambiguous though. On the one hand, there is evidence for positive mood facilitating safe driving [21]: Happy drivers caused fewer accidents. Zimasa et al. [19], however, could not find positive effects of joy on reaction time of hazard recognition in traffic.

1.2 Hypotheses

Results from experiments in the context of human-machine interaction indicate a correlation of emotion and user performance. Negative emotions like anger, sadness, frustration and anxiety are associated with high error rates, long reaction times, and accidents. Although literature suggests that positive emotions are associated with an improvement of performance, study results are mixed: Not all studies were able to show an increase of performance. Based on these insights, the hypotheses of the present study were formulated:

H1: High positive valence is associated with a low number of performance decrements.
H2: High negative valence is associated with a high number of performance decrements.

2 Methodology

2.1 Sample Description

Thirty employees of the Fraunhofer Institute for Communication, Information Processing and Ergonomics participated in the experiment. Eight data sets were excluded due to recording errors of the emotional valence. 45% of the remaining 22 subjects were female. The youngest participant was 17, the oldest 52 years old ($M_{age} = 30.96$; $SD_{age} = 9.76$).

2.2 Experimental Task

The experimental task from the anti-air warfare domain consisted of four subtasks.

1. Identify. Unidentified contacts had to be identified according to certain parameters and rules.
2. Warn. Hostile contacts entering the Identification Safety Range had to be warned.
3. Engage. Hostile contacts that entered the Weapon Range despite a warning had to be engaged.
4. NRTTs. "Non real-time tracks" (NRTTs) are contacts that the user had to add manually to the tactical display.

The present investigation was part of a larger experiment with several user states and multiple adaptation strategies tested. For the analysis of emotional valence, a ten minute interval from the control condition was selected. In the control condition, the interaction was not dynamically adjusted to the user state.

2.3 Operationalization of Variables

Performance. As a performance metric we used the number of incomplete tasks whenever at least one task was not completed within the assigned time limit. As long as time limits were not exceeded for any task present, a performance value of 0 was assigned to indicate adequate performance. Thus, a performance value (0 for adequate performance, or the current number of incomplete tasks) was available for every second of the investigation.

Emotional Valence. The construct of the emotional user state is operationalized by measuring the emotional valence in facial expressions using the emotion detection tool Emotient FACET (Fig. 1). FACET analyzes facial expressions in real time using a regular camera (in our case a Logitech C920 webcam). When a classification value exceeds the threshold of 0.5, the detected emotion is considered a moderate emotion [22]. Only two of the FACET classifiers, positive and negative valence, were used. The camera recorded 30 frames per second and generated classification results for every frame. Classification values were aggregated to calculate a second-by-second average.

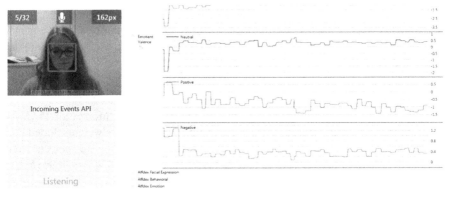

Fig. 1. Emotient FACET graphs for valence

2.4 Statistical Analysis

In order to avoid confounding within-subject factors and between-subject factors, regressions were calculated for every subject individually. Standardized regression coefficients of all subjects were averaged. A Fisher-Z-Test was used to test whether the mean differed significantly from 0.

3 Results

Overall means, standard deviations, minimal and maximal values for positive and negative emotional valence, and the number of incomplete tasks across the entire interval and all participants are presented in Table 1. Comparing the means of emotional valence shows a strong tendency towards negative valence. The overall mean of negative valence (0.595) is slightly higher than the threshold for an existing distinct emotion (0.5). The mean for positive valence, however, is in the negative range, indicating the absence of positive emotions. On average, the number of incomplete tasks was very low (0.775), indicating generally high performance.

Standardized regression coefficients including significance levels for all subjects are presented in Fig. 2. For positive valence (blue squares), a negative correlation was expected. For negative valence (orange triangles) a positive correlation was expected.

Table 1. Descriptive statistics of valence and number of performance decrements.

	M	*SD*	*Minimum*	*Maximum*
Positive emotional valence	−0.169	0.396	−0.999	0.999
Negative emotional valence	0.595	0.155	−0.807	0.998
Number of incomplete tasks	0.775	0.537	0	7

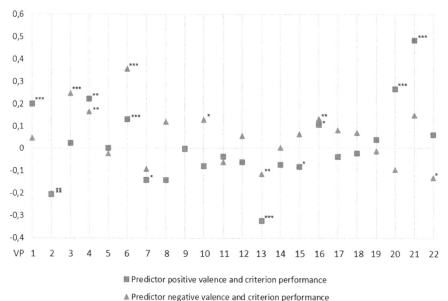

Fig. 2. Standardized regression coefficients of valence and performance for all subjects
(*** $p < .001$, ** $p < .01$, * $p < .05$) (Color figure online)

The correlation between positive emotional valence and the number of performance decrements can be described using the averaged standardized regression coefficient ($M = 0.016$, $SD = 0.174$), which did not differ significantly from 0 ($z = 0.02$, $p = .961$, $N = 22$). For 36% of the subjects, however, significant correlations between positive valence and performance decrements were found at the individual level. Some of these significant correlations were negative as expected. Others were positive, contradicting the hypothesis.

The averaged standardized regression coefficient for negative valence and performance decrements ($M = 0.030$, $SD = 0.136$) did not differ significantly from 0 ($z = -0.03$, $p = .923$, $N = 22$). Between negative valence and performance decrements, there were significant correlations for 45% of the subjects. Only some of these correlations were positive as expected.

4 Discussion

The present investigation aimed at gaining first insights into the benefit of adding the emotional user state to the diagnostic suite of an adaptive human-machine system. To that end, the correlations of emotional user states and performance in a C2 environment were examined. Based on theoretical background, positive valence was expected to improve performance (Hypothesis 1). It was assumed that negative valence impaired performance (Hypothesis 2). These correlations were investigated using regressions. The results are summarized in Table 2.

Table 2. Results of hypothesis testing

Hypothesis	Results
Hypothesis 1: A high positive valence is associated with a low number of performance decrements	On average, no significant correlations between positive valence and performance at the group level
Hypothesis 2: A high negative valence is associated with a high number of performance decrements	On average, no significant correlations between negative valence and performance at the group level

Results from this study did not match the findings of previous experiments and could not confirm the hypotheses. In several previous publications [6, 19–21], negative emotions were associated with impaired performance. In the present investigation, only some individuals showed this impairment. The improvement of performance associated with positive emotions discovered by James and Nahl [23], was not observed either. In an experiment with a similar C2 task, however, Panganiban [17] found effects that match the findings of the present investigation. There are several possible explanations for these results that are discussed below.

4.1 Individual Differences in Facial Expressions of Emotion

One possible explanation for the present findings are individual differences in the facial expressions of emotions. Cohn, Schmidt, Gross and Ekman [24] showed that facial expressions of feelings strongly vary individually but are stable over time. Only if subjects mirror their emotions in facial expressions, FACET is able to recognize them. Possibly, not all subjects showed their feelings sufficiently for automated classification based on facial expressions. In this case, an existing correlation between emotion and performance cannot be found empirically using facial expression diagnostics.

For some subjects, trends in the standardized regression coefficients were identified that indicate a correlation between emotional facial expression and performance. We found significant correlations between positive valence and performance for 36% and significant correlations between negative valence and performance for 45% of the subjects. These correlations did not always occur in the expected direction. This may explain why the overall means of the standardized regression coefficient were close to zero. It is possible that the emotional user state is a useful predictor of performance for some individuals, especially those who strongly mirror their emotions in facial expressions, but not for others. In this case, individual models of emotions would be helpful (cf. [25]).

4.2 Type of Tasks Unsuitable for the Investigation of Emotional States

Another possible explanation for the results is the type of task that was used. In general, emotional user states have rarely been investigated in C2 tasks and it is unclear whether and how the employed anti-air warfare task elicits emotional reactions. In the present experiment, emotional valence was on average neutral, only slightly negative. Positive states were infrequent, which is why they probably did not influence performance. Even though negative emotions were more common, the value for negative valence was only marginally higher than the threshold for a clear and distinct emotion. One possible reason is that the type of task used in this study did not induce high emotional valence. In this case, correlations between emotion and performance would be unlikely. Many other studies using the emotion detection software FACET present highly emotional stimuli. Examples are videos that aim at inducing emotional states [26], American presidential debates [27], or video games like Call of Duty [28].

Empirical evidence for the relationship of emotional user states and performance in C2 tasks was only found for the state anxiety, which was not associated with performance in a study by Panganiban [17]. She used an air defense task that showed similarities with the present experiment. Her subjects formed teams of two and coordinated fighter jets in an air battle. The mission was to protect civilians from hostile air contacts. Similar to the present investigation, neutral, less threatening, very threatening, and unidentified contacts had to be discriminated. Unidentified contacts, however, changed their identity autonomously and subjects were supposed to detect these changes. Contacts that identified as "very threatening" were detected more successfully than less threatening contacts. Panganiban [17] interpreted that the perceived threat narrows the attentional focus which left less threatening contacts less noticeable. The user state anxiety was not associated with change detection or performance in the defense task. However, subjects with a high trait anxiety showed higher performance in the latter.

Results and similarity of the task suggest that it is generally difficult to observe the correlation between emotional user state and performance in C2 tasks. Reasons for the absence of correlation may be found in general characteristics of C2 tasks. C2 is a military term, and execution of authority (command) and the surveillance of activities and goals (control) often occur in safety-critical areas. Errors can have severe consequences and should be avoided at all cost. Thus, the execution of a C2 task demands high concentration. Concentration can visibly influence facial expression [29]. The expression of concentration may have superimposed the emotional facial expressions. In that case, a correlation between emotion and performance might exist but cannot be observed in facial expressions. As subjective measures like questionnaires would regularly interrupt the task, they are not a viable alternative in a demanding operational context.

Another explanation could be the conscious or unconscious suppression of emotion to avoid distraction and facilitate concentration on relevant information. In that case, no correlation between emotion and performance would exist at all. Consequently, based on the available data, no clear statement can be made if a correlation truly exists or not. In order to examine the influence of emotions on performance in a C2 task, future studies can specifically induce emotional user states before task execution. This would allow statements about causality which cannot be drawn from the present data.

4.3 Methodical Reasons

Finally, methodical reasons may have contributed to the findings of this investigation. Experimental design and distribution of tasks in the scenarios were supposed to induce low performance. For most subjects, however, the induction did not succeed. The average number of incomplete tasks for each subject in both phases was only 0.7. This value indicates high overall performance which caused a ceiling effect: Episodes of low performance did not occur in sufficient frequency. Likewise, highly emotional states were rare. The overall high performance may have influenced the emotional state, leading to the absence of high emotional states like frustration or anger. It seems plausible that emotion and performance have a reciprocal influence on each other. Probably, the task was overall not demanding enough and therefore the data only reflected a small portion of the possible spectrum of performance and emotional valence. Low within-subject variance may have inhibited the expected statistical effects.

4.4   High Performance in Neutral Valence

If future studies were to show correlations between emotional states and performance in C2 tasks, the optimal emotional state for best possible performance could be assessed. Cai and Lin [30] searched for this "sweet spot" in a driving simulator study. They found an inverted u-shaped relationship between performance and both valence and arousal (Fig. 3), meaning that performance drops with high valence or high arousal (cold colors in Fig. 3). Performance is highest with neutral valence and medium arousal (warm colors in Fig. 3). These results represent an alternative explanation model for the findings of the present study. However, in our investigation, the induction of low performance did not succeed. Further studies that cover the entire spectrum of performance and emotions are necessary to determine the sweet spot in the C2 domain. In a future adaptive system, deviations from this "sweet spot state" could be determined and addressed by suitable adaptations to mitigate performance decrements caused by high emotional valence.

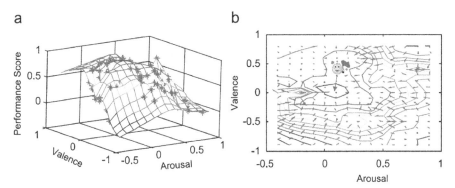

Fig. 3. Inverted u-shaped relationship between performance, valence and arousal. Image taken from Cai and Lin [30] (Color figure online).

5   Conclusions

Practical implications for future studies and the design of an adaptive system can be drawn from the gained results.

5.1 Practical Implications

In future studies, the methodology should be revised considering following aspects:

1. Increasing task density should generate higher variance on subject level in performance and possibly also in emotional user states. Inducing a larger emotional spectrum would enable the assessment of the emotional sweet spot for best possible performance (cf. Cai and Lin [30]) in C2 environments.
2. Data should be collected from a larger sample, as technical difficulties with emotion detection tools can reduce sample size.
3. Other measures of the emotional state could be used in addition to Emotient FACET. Another non-intrusive emotion detection tools is, for example, AFFDEX by Affectiva [31].
4. Based on Russell's Circumplex model of affect [32], emotion "can be understood as a linear combination of two dimensions, [...] valence and arousal" [33, p. 716]. In order to cover a broader emotional spectrum, both dimensions could be assessed. Heart rate variability may serve as an indicator for arousal [34].
5. Even though a significant correlation between emotional valence and performance was found for several subjects, this investigation does not allow any statement about causality. To examine cause and effect, the induction of emotional user states before task execution is necessary. For example, performance in positive states could be compared to performance in neutral and negative states.
6. The expression of emotions varies considerably among participants. Therefore, individual models of the emotional user state should be considered in future adaptive systems.

The present work is a pre-investigation for the integration of the emotional user state in an adaptive system developed by Fuchs et al. [5]. Based on the results, it can merely be concluded that performance during low work load is not associated with emotional valence is the employed C2 task. Whether or not this is also true in high work load/low performance scenarios, should be examined in future studies. Results were consistent with Cai and Lin [30] who found that performance was highest with neutral valence. To find their proposed emotional sweet spot, further investigations are necessary that cover a broader spectrum of performance and emotion. If the neutral emotional state can be confirmed as a sweet spot, adaptation strategies could aim at shifting the emotional state to that spot in case of a performance decrement.

References

1. Bainbridge, L.: Ironies of automation. Automatica **19**(6), 775–779 (1983)
2. Parasuraman, R., Manzey, D.: Complacency and bias in human use of automation: an attentional integration. Hum. Factors **52**(3), 381–410 (2010)
3. Åkerstedt, T., Mollard, R., Samel, A., Simons, M., Spencer, M.: The role of EU FTL legislation in reducing cumulative fatigue in civil aviation. Presented at the ETSC Meeting (2003)
4. Young, M.S., Stanton, N.A.: Malleable attentional resources theory: a new explanation for the effects of mental underload on performance. Hum. Factors: J. Hum. Factors Ergon. Soc. **44**(3), 365–375 (2002)

5. Fuchs, S., et al.: Anwendungsorientierte Realisierung adaptiver Mensch-Maschine-Interaktion für Sicherheitskritische Systeme (ARAMIS). Final project report (grant no. E/E4BX/HA031/CF215). Fraunhofer FKIE, Wachtberg (2019)

6. Eyben, F., et al.: Emotion on the road-necessity, acceptance, and feasibility of affective computing in the car. Adv. Hum.-Comput. Interact. **2010**(1768), 1–17 (2010)

7. Jones, C.M., Jonsson, I.-M.: Automatic recognition of affective cues in the speech of car drivers to allow appropriate responses. In: Proceedings of the 17th Australian Conference on Computer-Human Interaction, pp. 1–10. Citizens Online: Considerations for Today and the Future, Canberra (2005)

8. Groeger, J.A.: Understanding Driving: Applying Cognitive Psychology to a Complex Everyday Task. Taylor & Francis Ltd., Hove (2000)

9. Kontogiannis, T.: Stress and operator decision making in coping with emergencies. Int. J. Hum.-Comput. Stud. **45**(1), 75–104 (1996)

10. Madhavan, P., Wiegmann, D.A., Lacson, F.C.: Automation failures on tasks easily performed by operators undermine trust in automated aids. Hum. Factors **48**(2), 241–256 (2006)

11. Vassiliou, M.S., Alberts, D.S., Agre, J.R.: C2 Re-Envisioned: The Future of the Enterprise. CRC Press, Boca Raton (2014)

12. Littlewort, G., et al.: The computer expression recognition toolbox (CERT). In: Proceedings of the 2011 International Conference on Automatic Face & Gesture Recognition and Workshops, pp. 298–305, (2011)

13. Lemay Center for Doctrine: Key considerations of command and control. Annex 3-30 - Command and Control, pp. 6–10 (2014)

14. Vassiliou, M.S., Alberts, D.S., Agre, J.R., Alberts, D.S., Agre, J.R.: C2 Re-Envisioned: The Future of the Enterprise. CRC Press, Boca Raton (2014)

15. Wohl, J.G.: Force management decision requirements for air force tactical command and control. IEEE Trans. Syst. Man Cybern. **11**(9), 618–639 (1981)

16. Gaillard, A.: Concentration, stress and performance. In: Hancock, P.A., Szalma, J.L. (eds.) Performance Under Stress, pp. 59–65. Ashgate Publishing, Burlington (2008)

17. Panganiban, A.: Effects of anxiety on change detection in a command and control task. Master of Science thesis, University of Cincinnati, Cincinnati (2011)

18. Stecklov, G., Goldstein, J.: Terror attacks influence driving behavior in Israel. Proc. Nat. Acad. Sci. (PNAS) **101**(40), 14551–14556 (2004)

19. Zimasa, T., Jamson, S., Henson, B.: Are happy drivers safer drivers? Evidence from hazard response times and eye tracking data. Transp. Res. Part F: Traffic Psychol. Behav. **46**(1), 14–23 (2017)

20. Wells-Parker, E., et al.: An exploratory study of the relationship between road rage and crash experience in a representative sample of US drivers. Accid. Anal. Prev. **34**(3), 271–278 (2002)

21. Deffenbacher, J.L., Oetting, E.R., Lynch, R.S.: Development of a driving anger scale. Psychol. Rep. **74**(1), 83–91 (1994)

22. iMotions: iMotions facial expressions analysis pocket guide (2016)

23. James, L., Nahl, D.: Road Rage and Aggressive Driving: Steering Clear of Highway Warfare. Prometheus Books, Amherst (2000)

24. Cohn, J.F., Schmidt, K., Gross, R., Ekman, P.: Individual differences in facial expression: Stability over time, relation to self-reported emotion, and ability to inform person identification. In: Proceedings of the 4th IEEE International Conference on Multimodal Interfaces, pp. 491–496 (2002)

25. Hudlicka, E.: Affect-adaptive interaction in behavioral health technologies. In: Proceedings of the 22nd International Conference on Human-Computer Interaction in Copenhagen, in press (2020)

26. Taggart, R.W., Dressler, M., Kumar, P., Khan, S., Coppola, J.F.: Determining emotions via facial expression analysis software. Behav. Res. Methods **50**(4), 1446–1460 (2016)

27. Fridkin, K.L., Gershon, S.A., Courey, J., LaPlant, K.: Gender differences in emotional reactions to the first 2016 presidential debate. Polit. Behav., 1–31 (2019). https://doi.org/10.1007/s11109-019-09546-9
28. Iqbal, F.: Understanding user interaction in a video game by using eye tracking and facial expressions analysis. Unpublished Master's thesis, University of Tampere, Tampere, Finnland (2015)
29. Takahashi, K., Yamada, K., Nakano, T., Yamamoto, S.: Method of detecting concentration on cellular phone call from facial expression change by image processing. In: 2005 IEEE International Conference on Systems, Man and Cybernetics, vol. 4, pp. 3444–3448 (2005)
30. Cai, H., Lin, Y.: Modeling of operators' emotion and task performance in a virtual driving environment. Int. J. Hum.-Comput. Stud. **69**(9), 571–586 (2011)
31. McDuff, D., Kaliouby, R., Kassam, K. Picard, R.: Affect valence inference from facial action unit spectrograms. In 2010 IEEE Computer Society Conference on Computer Vision and Pattern Recognition - Workshops, pp. 17–24 (2010)
32. Russell, J.: A circumplex model of affect. J. Personal. Soc. Psychol. **39**, 1161–1178 (1980)
33. Posner, J., Russell, J.A., Peterson, B.S.: The circumplex model of affect: an integrative approach to affective neuroscience, cognitive development, and psychopathology. Dev. Psychopathol. **17**(3), 715–734 (2005)
34. Appelhans, B.M., Luecken, L.J.: Heart rate variability as an index of regulated emotional responding. Rev. Gen. Psychol. **10**(3), 229–240 (2006)

The Platonic-Freudian Model of Mind: Defining "Self" and "Other" as Psychoinformatic Primitives

Suraj Sood[✉]

The Sirius Project, Scarborough, UK
thesiriusproj@gmail.com

Abstract. As "nurtural" (rather than merely natural) kinds of human beings, people are complex and multifaceted. Any complete human science would require a complete theory of persons. Accomplishing the latter is the core objective of the present article.

First, a feature list first laid out in [1] is summarized. This list is briefly critiqued. Next, the concept of *person* engaged with is expanded with the addition of nine novel features. These features follow from "holarchic psychoinformatics" [2], which was first propounded as a step forward from Sood's analytic treatment of third-force, existential-humanistic psychology. Person is formalized as a function of self and other; they are also granted to be romantic, existential, humanistic, chemical, environmental, hedonic and eudaimonic (happiness-seeking), conservative, and liberal. These are in addition to persons being physical, biological, psychological, social, cultural, and spiritual. Sood's holarchic view of persons is enlarged.

Psychologically, augmented cognition as an established field of research and practice begets the formal studies of augmented affect, augmented behavior, and augmented motivation. All such interdisciplinary fields are required for the human-computer interactionist's study of augmented mind, more broadly.

Additionally, this article builds on the person-situation interaction framework formalized in [1]. It does so by adding a formalization following from the discussion of psychological situations put forward by Rauthmann, Sherman, and Funder in [3]. The formalization of psychological situations sets them as a function of *cues*, *characteristics*, and *classes*. Further psychological equations that follow from this article's formalisms of person and situation, when considered along with Sood's formulae for mind and behavior, are then presented.

Keywords: Holarchy · Person-situation interaction · Third-force psychology · Self · Other

The original version of this chapter was revised: Section 4 has been modified. The correction to this chapter is available at https://doi.org/10.1007/978-3-030-50353-6_22

D. D. Schmorrow and C. M. Fidopiastis (Eds.): HCII 2020, LNAI 12196, pp. 76–93, 2020.
https://doi.org/10.1007/978-3-030-50353-6_6

1 Introduction

In John Carpenter's 1974 movie Dark Star, Commander Powell instructs spaceship commanding officer Doolittle to teach a robotic bomb "a little phenomenology" [46]. Doolittle then engages in Socratic discourse with Bomb #20, hoping to prevent it from fulfilling its singular task: to explode. The discourse includes a Cartesian meditation on proper skepticism toward the sensory data taken in by a given being. Based on this, Doolittle argues that Bomb #20 cannot prove that its command to explode is working data; Bomb #20 seems to accept this, but extends the logic to include Sergeant Pinback as being "false data". Since it would be illogical to act upon faulty data, Bomb #20 tunes Pinback out in favor of fulfilling its purpose. Bomb #20 detonates.

The interactions between Doolittle and Pinback with Bomb #20 impart an important lesson for human-computer interaction (HCI). Specifically: we must communicate logically with robots, or else they might act against their human creators' best wishes. Thus, it behooves HCI and the augmented cognition community to understand humans, computers (e.g., robots), and how they interact in order to augment their collective, cognitive phenomenology.

In this article, a framework of persons presented in [1] is critiqued and subsequently expanded. Following this, the concept of *person* engaged with is expanded with the addition of nine novel features. These features follow from Sood's [2] "holarchic psychoinformatics", which was first propounded as a step forward from his analytic treatment of third-force, existential-humanistic psychology. *Person* is formalized as a function of self and other; they are further granted to be romantic, existential, humanistic, chemical, environmental, hedonic and eudaimonic (happiness-seeking), conservative, and liberal. These are in addition to persons being physical, chemical, biological, psychological, social, cultural, and spiritual (as explicated in Sood's [2] theory of reality). Put more simply: The holarchic view of persons is enlarged.

Additionally, this article builds on the person-situation interaction framework formalized in Sood [1]. It does so by adding a formalization following from the discussion of psychological situations put forward by Rauthmann, Sherman, and Funder in [3]. The formalization of psychological situations sets them as a function of *cues*, *characteristics*, and *classes*. Further psychological equations that follow from this article's formalisms of person and situation, when considered along with Sood's formulae for mind and behavior, are presented. A series of novel concepts relevant to augmented cognition as a function of HCI are then propounded. In the last place, computational psychology is discussed in relation to theoretical psychology in the service of strengthening human-machine symbiosis.

2 Related Work

Sood [20] posited 18 human features in answering the question of what it means to be human (or, perhaps to be more specific—*people*). Sood's features—with the additions of sub-features *learning* and *attention* within feature 18—included

1. *Physical* – People's bodies are composed of matter. Further, people interact with other physical objects.

2. *Biological* – People breathe, eat, and drink; and a great many of them have sex and reproduce.
3. *Temporal* – People are born, they live, and they die; they experience time.
4. *Cultural* – People are embedded in cultures characterized by unique but shared ways of being.
5. *Social* – People participate in societies consisting of concrete relations between themselves and others.
6. *Economic* – People are agents who trade goods and services with one another in marketplaces.
7. *Technological* – People invent and utilize tools to perform tasks they were previously unable or less able to accomplish.
8. *Artistic* – People express themselves through the creation of original works such as paintings and songs.
9. *Intellectual* – People aim to comprehend reality and achieve accurate understandings of it.
10. *Moral* – People have unique and shared ideas of wrong versus right action.
11. *Spiritual* – People seek enlightenment, wisdom, and contact with the divine or supernatural via practices such as meditation and prayer.
12. *Religious* – People worship what they deem as sacred (e.g., God or Gods) through rituals and organized communion.
13. *Political* – People negotiate and have interests that are in line or at odds with those of others.
14. *Athletic* – Whether for fitness or organized play, people exercise their bodies and minds.
15. *Professional* – People work toward particular goals, including earning money and achieving satisfaction.
16. *Recreational* – People enjoy leisurely activities such as taking walks and attending parties.
17. *Linguistic* – People communicate via representational symbol systems characterized by semantics, syntax, and pragmatics.
18. *Psychological* – People have minds and engage in behaviors. More specifically, they think, feel, have personalities, interact with situations, are motivated, sense, perceive, experience, learn, and pay attention.

The above list may be considered more relevant to personology than personality psychology. McCrae and Costa [22] discussed "personologists" (p. 81) but did not distinguish such researchers from personality psychologists. Such a distinction should be worked out for technical reasons: primarily, that of precision. Modern personality psychology is defined as involving: 1) "an emphasis on defining and understanding individual differences"; and 2) "an emphasis on the ways in which the various parts of the person are organized" ([23], p. 100). Kukla [47] states that the personality theorist's task is to put together the pieces of human nature independently constructed by psychology's "process areas" (including social and perceptual psychologies) (p. vii). Personology is simply the study of persons.

Given list item 18 which states the core topics of psychological inquiry, one could most reasonably expect E-H psychologists to focus on the sum-total of its items. These psychologists need to include cognition, affect, personality, situationality, behavior, motivation, sensation, perception, experience, learning, and attention in their ultimate descriptions of who people are, their explanations of how people come to be, and their predictions of whom people are expected to become.

Within psychology, Freud was the pioneer of personality vis-à-vis mind as much as Skinner was the same for behavior [13]. Affect has been addressed by psychologists via the five factor model (FFM) constructs of Extraversion (positive emotion) and Neuroticism (negative emotion); cognition was included in Kelly's [28] personal construct and Dweck and Leggett's [29] social-cognitive theories. Experience, meaning, and motivation have been taken up by third force theorists such as Kelly, Maslow, and Rogers, in addition to positive psychologists (see, e.g., Proctor, Tweed, and Morris [30]). Lastly, learning has been covered by Bandura and Huston [51] and attention by philosophical, perceptual, and cognitive psychologists (e.g., William James).

Despite the progress summarized above, it remains an open question whether psychologists have fully accounted for both people and their situations. What determines their interaction? The best-established construct that is closest to the former is personality. Situations, on the other hand, have no corresponding construct denoting situationality. It may be partially inferred from this latter fact that psychologists understand personality better than situationality. Formalizing *situation* is undertaken later in this paper. First, however, Sood's multidimensional notion of *person* is expanded before being formalized in terms of more basic, psychological primitives. Such primitives have been introduced in this and previous works to address what has been called "the units of analysis problem in psychology" ([31], p. 177).

3 Expanding the "Person" Concept

3.1 Romantic, Existential, Humanistic, Chemical, Environmental, Hedonic and Eudaimonic, Conservative, and Liberal Features

The following nine features are now added to Sood's original list:

19. *Romantic* – People become emotionally involved with one another. Most get married.
20. *Existential* – People are responsible and free: they have "psychological wills"[1].[2]
21. *Humanistic* – People are creative, spontaneous, and active beings who contribute to the furthering of humanity.
22. *Chemical* – People are composed in part of physical reactions taking place throughout their bodies.
23. *Environmental* – People engage in a variety of ways with their surroundings.

---

[1] Various theories of psychological will have been proposed in the past two centuries, ranging from Friedrich Nietzsche's "will to power" to Viktor Frankl's "will to meaning". Will is used in the present context to refer simply to volition, i.e. purposive striving evidenced when one or more individuals decides on and commits to certain action.

[2] This article's notion of existential humanness is intended to be fully compatible with Yalom's existential givens, including freedom, death, isolation, and meaning [45].

24. *Hedonic* – People seek happiness in the form of pleasure.
25. *Eudaimonic* – People seek happiness in the form of fulfillment.
26. *Conservative* – People live in accordance with rules and principles designed with security in mind.
27. *Liberal* – People live freely to maximize (e.g.) diversity, inclusion, and peace.

19 is added since it is arguably not fully subsumable beneath 5 (*social*); nor, probably, within 9 (*intellectual*) (contra-cognitivist models that view affect as merely a class of cognition: see [10] for a treatment of this perspective). Despite this, marriage is an established social and religious (12) convention. Moreover, romantic being in the present sense is mostly meant as being affective (in the same sense as in 18: people "feel"). It could be expressed artistically (8), but is not reducible to such. 19 could thereby be viewed as a product of 5, 8, 9, 12, and 18, though it need not be necessarily. 26 and 27 are meant not merely in their political senses, but rather more broadly to encompass human being and doing.

People are also inherently humanistic and environmental. For the theoretical psychologist, the latter is to be sharply distinguished from persons' being situational as in 18. Varela et al. [11] propounded the original enactive framework marrying cognitive science with phenomenology, asserting that "the organism both initiates and is shaped by the environment" (p. 174). Sood [1] substituted "organism" and "environment" in this statement with "person" and "situation" respectively, asserting that doing so rendered his treatment more psychological (and so less biological). Lewin [12] formalized human behavior as

$$Be = F[P, E] \qquad (1)$$

Where *Be* equaled "behavior", *P* equals "person", and *E* equals "environment". Lewin's statement reads "Behavior = Function of person and environment" (p. 878). Sood's [2] formalization of human behavior was assigned the variable *B*, which equaled $F[Sm, Rp]$

$$B = F[Sm, Rp] \qquad (2)$$

Where *Sm* equals "stimulus" and *Rp* equals "response". Sood revised Lewin's behavioral formula to align more with traditional behavioristic psychology—viz., Skinner's [13]—and truncated the latter's *Be* variable to simply *B*.

Sm could be regarded as analogous, if not identical, with *E* in that for behaviorists like Skinner, stimuli were objects in the subject's surrounding (external) environment. For Rauthmann et al. [3], situations are composed partially of environmental *cues* that are physical, objectively quantifiable stimuli. While this framing suggests that the environment should be conceived as being part of situationality, for the present discussion, the interchangeability of *E*, *Sm*, and situationality (formalized later in this chapter) is simply to be regarded as noteworthy.

3.2   The Person Equation

Psychological persons may be said to be either selves or others. Psychological notions of *self* and *other* pervade the existential psychological literature [4; for Martin Buber's discussion of the related "I and thou" phenomenon, see 5]. Sood [1] formalized person-situation interaction as a complex, interdependent function of mental and behavioral structures (i.e., states and traits) and processes. He did so as follows[3,4]

$$[P, S] = F\big[St_{(T,Se)}, Pc\big]_{(M,B)} \tag{3}$$

Where equals "person", equals "situation", equals "structure", equals "trait", equals "state", equals "process", M equals "mind", and equals "behavior" [20]. (Traits and states are treated as distinct types of psychological structures.) $[P, S]$ is a whole composed entirely of parts St_{MSe}, St_{BSe}, St_{MT}, St_{BT}, Pc_M, and Pc_B, which respectively denote "mental states", "behavioral states", "mental traits", "behavioral traits", "mental processes", and "behavioral processes". Informally, (3) reads: *person-situation interaction is a function of mental and behavioral structures[5] and processes.*

The more direct formalization of P is now undertaken

$$P = F[Sl, Ot] \tag{4}$$

Where Sl equals "self" and Ot equals "other". Self has received a recent psychological treatment in Klein [7], where William James' classical notions of "self-as-known" and "self-as-knower" received updates to a more holarchic notion. Klein identifies two distinct kinds of selves from cognitive neuroscience and clinical case work involving memory and knowledge, in particular. Klein's epistemological self is "the self of neural instantiation: the neuro-cognitive categories of self-knowledge" (p. 20); his ontological self is "the self of first-person subjectivity…that consciously apprehends the content of the epistemological self" (p. 46). Klein's dualistic view of self may be characterized as holarchic—i.e., subjective-objective—to the extent that neuro-cognition is ontologically objective (in Searle's sense [8]) whereas first-person subjectivity (as discussed topically in, viz., [9]) is ontologically subjective. Dennett [17] also offered a novel theory of self, defining it as a center of narrative gravity, a "purely abstract object…[and] fiction". This treats self entirely as an ontologically subjective phenomenon.

Other is a concept that, while theoretically opposed from the notion of self, has received marginal attention from psychologists. The possibilities of the psychological study of other open numerous such constructs. Hyphenated constructs for the same study of self include: self-compassion, self-confidence, self-control, self-distancing, self-doubt, self-efficacy, self-expansion, self-harm, self-reflection, self-suppression [44], self-determination, self-care, self-loathing, self-comparison, self-concept, self-esteem,

---

[3] All sufficiently similar equations offered hereon are syntactically consistently with Lewin's field theory of behavior [12] and Sood's enactive person-situation formula. Two-letter variable-naming is allowed to the extent that the same is in software program variable declaration, and is particularly necessary in cases of multiple constructs beginning with identical first letters.

[4] The kind of person-situation interaction expressed through (4)—which has been formalized to render the construct more applicable within mathematical, theoretical, and computational contexts—is thus distinctly psychological in accommodating mind and behavior. Mind, behavior, person, and situation comprise modern psychology's highest-level topics of study.

[5] Such mental and behavioral structures may be either traits, states, or hybrid "trates" (this appears to be a novel neologism). The use of *trate* over both *state* and *trait* would eliminate the need to use latter in formalisms such as (3).

self-handicapping, self-image, self-perception, self-regulation, self-reference ([40], p. 253), self-actualization [41], self-transcendence, self-knowledge, self-ignorance ([39], p. 713), self-interest, self-report, self-replication (as denoted by "autopoiesis"—see [43] for a discussion of autopoietic technologies), self-directed, self-talk, self-reliance, self-realization, self-defeating, self-concept, self-identify, self-as-known, self-as-knower, self-sabotage, self-aggrandizement, self-effacement, self-evident, self-love, and self-consciousness. Thus, potential constructs for the human study of other include: *other-compassion, other-confidence, other-control, other-distancing, other-doubt, other-efficacy, other-expansion, other-harm, other-reflection, other-suppression, other-determination, other-care, other-loathing, other-comparison, other-concept, other-esteem, other-handicapping, other-image, other-perception, other-regulation, other-reference, other-actualization, other-transcendence, other-knowledge, other-ignorance, other-interest*[6]*, other-report* ([2], p. 352), *other-replication, other-directed, other-talk, other-reliance, other-realization, other-defeating, other-concept, other-identify, other-as-known, other-as-knower, other-sabotage, other-aggrandizement, other-effacement, other-evident, other-love,* and *other-consciousness.* Further "other" concepts likely remain to be named.

In the context of augmented cognition, the central role of the self who thinks (i.e., from Descartes' original *cogito*) is easily imagined and difficult, if not impossible, to successfully refute. However, the notion of "other minds" continues to plague the analytic philosophy of mind, in particular (see the famous "problem of other minds" [14]). Resolving this problem, which consists in answering whether—and, if so, then how—we may come to know that other people's minds exist, would be tantamount to knowing how we could know of other people's cognitions (and affects and motivations, a la Sood's Platonic-Freudian formula of mind below). Such knowledge would be requisite for its instrumentalization; and, if technology consists in the instrumentalization of knowledge—not merely of information, which is truth value-neutral—then knowledge of other minds is requisite for any technology that would augment user cognition.

Knowledge here is understood as a form of cognition represented in one's mind, or enacted procedurally via the skilled use of one's body. In either case, it is encoded in an "embodied", neurocognitive substrate. Representing knowledge via formal syntax and operations is the domain of mathematical logic, which also extends into computation primarily in the form of discrete logical operations determined by the programmer.[7]

3.3   The Platonic-Freudian Model of Mind

Sood [2] formalized *M mind* as a portion of ψ, *psychology* (denoting the field of psychology, including its two most eminent and high-level, modern topics of study). He did so as follows

$$\psi = F[M, B] \tag{5}$$

---

[6] Other-interest has been covered by Gerbasi and Prentice in their creation of the Self- and Other-Interest Inventory [42].

[7] See [15] for a critique of the embodied view of knowledge, and [16] for a systematic illustration of the (computational) cognitivist approach to it.

Where B equals *behavior*. Sood's formalism was disciplinary in nature; person-situation interaction (with (M, B) operating as a subscript to $\left[St_{(T,Se)}, Pc\right]$) could just as easily have been set equal to ψ. In any case, M was next formalized as follows[8]

$$M = F\left[(A, C, Mv)_{(U-,Sb-)Cs}\right] \qquad (6)$$

Where A equals "affect", C equals "cognition", Mv equals "motivation", $U-$ equals "un-", $Sb-$ equals "sub-", and Cs equals "consciousness". According to the right portion of (6)'s subscript, each of these elementary mental phenomena may be either unconscious, subconscious, or conscious. (8) yields the following nine constructs: "unconscious affect", "subconscious affect", and "conscious affect"; "unconscious cognition", "subconscious cognition", and "conscious cognition"; and "unconscious motivation", "subconscious motivation", and "conscious motivation". Any of these constructs could informally be considered *subminds* in a manner analogous to how each of the five-factor model's traits (Openness, Conscientiousness, Extraversion, Agreeableness, and Neuroticism) could be regarded as *subpersonalities* of a given person's overall personality.

The fusion of (2), (3), and (6) is now undertaken

$$[P, S] = F\left[St_{(T,Se)}, Pc\right]_{[(A,C,Mv,B)_{(U-,Sb-)Cs}]} \qquad (7)$$

(7) reads: Person-situation interaction is a function of unconscious, subconscious, and conscious affective, cognitive, motivational, and behavioral traits, states, and processes. B here was not part of Sood's original Platonic-Freudian model of mind, nor does (7) need to imply that it now is. It has been added to (7) (and subsequent invocations of $F\left[St_{(T,Se)}, Pc\right]$) given its historical closeness with Mv (see the conation concept, as in Plato's work [37]), and to accommodate Maslow's view of behavior almost always requiring motivation in order to occur.

The tripart primitives listed in Table 1 follow from the compound psychological primitives named between (6) and (7), as well as from (7)

Table 1. Tripart primitives of Sood's Platonic-Freudian model of mind (References to constructs of the five-factor model of personality (a.k.a. the "Big Five") are derived from [35].)

Structures and Processes	Freudian topography		
Platonic triad (including Mv instead of B)	*CsASe*: Conscious affective state (e.g., palpable moods; feelings "of the moment")	*CsMvSe*: Conscious motivational state (e.g., realizing to reach a goal state or overcome a given situation)	*CsCT*: Conscious cognitive trait (e.g.: "woke", i.e., subjectively self-aware *Openness*")

(continued)

---

[8] Formula (6) drew from Freud's topographical model of mind [18] on one hand—where mental content passes between the unconscious and conscious sub-minds via the intermediary subconscious—and Revelle's recent attempt to synthesize Plato's tripartite model of mind (consisting of precursors for affect, cognition, and motivation) into a formal personality framework [19].

Table 1. (*continued*)

Structures and Processes	Freudian topography		
	CsAT: Conscious affective trait (e.g., "woke" *Neuroticism*)	*CsMvT*: Conscious motivational trait (e.g., "woke" *Conscientiousness*)	*CsCSe*: Conscious cognitive state (e.g.: being pensive; being momentarily lost or absorbed in thought, i.e. introspective, reflective, ruminative, imaginative, cogitative; daydreaming)
	SbCsCSe: Subconscious cognitive state (e.g., REM-dreaming)	*SbCsASe*: Subconscious affective state (half-awareness of mood, fleeting feelings or emotionality)	*SbCsMvSe*: Subconscious motivational state
	SbCsCT: Subconscious cognitive trait (e.g., Jungian/Myersian "iNtuitive" type)	*SbCsAT*: Subconscious affective trait (half-awareness of *Neuroticism* or *Extraversion – Enthusiasm*)	*SbCsMvT*: Subconscious motivational trait
	UCsMvSe: Unconscious motivational state	*UCsASe*: Unconscious affective state (e.g., subject undergoing intuitive processing)	*UCsCSe*: Unconscious cognitive state (e.g., subject undergoing intuitive processing)
	UCsMvT: Unconscious motivational trait (e.g., *Conscientiousness – Industriousness*)	*UCsAT*: Unconscious affective trait	*UCsCT*: Unconscious cognitive trait (e.g., *Openness – Intellect* or *Imagination*)
	CsAPc: Conscious affective process (e.g., processing of emotions during appropriate psychotherapeutic intervention)	*CsMvPc*: Conscious motivational process (setting one's mind to accomplish a goal or complete a task)	*CsCP*: Conscious cognitive process

(continued)

Table 1. (*continued*)

Structures and Processes	Freudian topography		
	SbCsCPc: Subconscious cognitive process	*SbCsAPc*: Subconscious affective process	*SbCsMvPc*: Subconscious motivational process
	UCsMvPc: Unconscious motivational process	*UCsAPc*: Unconscious affective process (e.g., System 1 intuition)	*UCsCPc*: Unconscious cognitive process (e.g., System 1 intuition)

The mathematical-theoretic approach to psychology undertaken here and in [1, 2] requires an expansion of Freudian's topography. Specifically, if unconsciousness is zero (0)-awareness; subconsciousness is half (0.5)-awareness; and consciousness is full (1)-awareness, then mathematically, one could speak equally of negative subconsciousness and consciousness (-0.5-awareness and -1-awareness, respectively). Negative psychology is to be contrasted with positive psychology in that the former includes psychopathology and clinical and abnormal psychologies, whereas positive psychology's core topic of study is well-being.

Sood [1, 2] formalized the sub-primitives of Table 1's triads into his person-situation and mind formulae, but neglected to explicate them rigorously. *Affect* has been defined in [24] as "a non-conscious experience of intensity...a moment of unformed and unstructured potential". It "cannot be fully realized in language" and "is always prior to and/or outside of consciousness". It is "the body's way of preparing itself for action in a given circumstance by adding a quantitative dimension of intensity to the quality of an experience".

Starting with Kahneman and Tversky's pioneering work (laid out for a more mainstream audience in [25]), *cognition* in modern psychology has frequently been defined as consisting of Systems 1 and 2. In this "dual-process" theoretic model, System 1 consists of thought that is fast, instinctive, affective, and unconscious[9]. System 2, on the other hand, consists of slower, more deliberative, logical, and conscious cognition[10].

Motivation has been defined differently by varying theorists. The basic question for the science of motivation is why beings (viz.: humans, animals, and/or robots) do what they do. Anthropological theories of motivation abound. In psychology, prominent such theories include drive-reduction, evolutionary, and optimal arousal theories. A parsimonious, complete theory of motivation would minimally need to answer the questions of why such beings want to, should, need to, and simply do carry out their behaviors.

Maslow developed the now-famous hierarchy of needs in which humans successively fulfill needs of varying classes [27]. He believed that motivations could be meaningfully

---

[9] Intuition is an unconscious process, but its outputs (occurring as insight, e.g. realized by a subject in an "aha!" moment) are conscious.

[10] See [26] for a further unpacking of this framing.

separated into groups based on two criteria: 1) which of them must be acted upon first in order for a person to survive, and 2) which are necessary to act upon for attaining "self-actualization" (pp. 375–382). Maslow and Horney [36] considered self-actualization respectively as a syndrome and trait of neurotic personalities, reflecting the state of psychological theory during the mid-20th century as psychoanalytic and clinical more so than "positive" (i.e., being more interested in human growth and potential, flourishing, and well-being).

In decreasing order of their relative degrees of necessary fulfillment, Maslow's motivations were "physiological", "safety", "love", "esteem", and self-actualization needs. Still, the question of *how* motivations such as these interact—both with one another and with other factors (cognitive, emotional, behavioral)—has not yet been answered. For Maslow, motivation was almost always necessary for behavior; additionally, he believed that more than one motivation typically figures into a single behavior (p. 370). It would seem, therefore, that a Maslowian science of motivation would need to work first from behavior to motivation, and possibly only afterward toward other psychological phenomena (e.g., affect and cognition).

Maslow [27] stated that "*any* conscious desires (partial goals) are more or less important as they are more or less close to the basic needs [of Maslow's hierarchy]" (p. 384). If it is accepted that Maslow recognized some fundamental connection between desires and goals, then it would follow from an earlier statement he made—i.e., that "conscious, specific, local-cultural desires are not as fundamental in motivation theory as the more basic, unconscious goals" (p. 370)—then for Maslow, unconscious wants will always be closer to our basic needs than will conscious ones[11].

Kelly [28] considered the following to be motivational concepts: "laziness", appetite, and affection (p. 77). Kelly understood motivations as being parts of greater systems of construction evidenced by individuals (e.g., through dialogue). Considered this way, motives can be thought to play out within close proximity to beliefs. However, "appetite" is closer to one of Maslow's basic needs, and is also something of a raw instinctual property characterizing a living being's consumptive capacity with respect to a suitable object Y (e.g., food). Meanwhile, Kelly himself provided another, more historical framing of motivation in terms of the synonymous triads of "cognition, *conation* [emphasis added], and affection", "intellect, *will* [emphasis added], and emotion", and (in more modern terms) "thought, *action* [emphasis added], and feeling" (pp. 68–69). Motivation, then—to the extent that it is will manifest in one's action—may be determined as such in reverse fashion from behaviors. However, Maslow [27] asserted that "Motivation theory is not synonymous with behavior theory... While behavior is almost always motivated, it is almost always biologically, culturally, and situationally determined as

---

[11] The theory that unconsciousness is "greater" in various respects (more influential motivationally, for the present context) than consciousness goes back to at least Freud; this theory may be more amenable to a truly scientific analysis today. Regardless, there remain the scientific questions associated with motivation's operationalization. Two such questions may be posed. First: How do we identify distinct motivations as such—both in terms of their relatively more autonomous properties, and their interactions in a person or persons' overall motivational system(s)? And second—how do motivations interact with other psychological phenomena like cognitions and emotions?

well" (p. 371)[12]. Regarding the possibility of *un*motivated behavior, Maslow stated that "expressive behavior is either unmotivated or…less motivated than coping behavior" ([38], p. 138). Expressive behavior is unconscious (*UCsB*) while coping behavior is conscious (*CsB*).

States and traits could be viewed as being conditioned respectively by situations and persons. A more mathematical view may be taken, whereby states and traits are distinguishably measurable based on a psychological phenomena's duration or degree of stability vs. plasticity. It is presumed that states would be of lesser duration and greater stability, while traits are of greater duration and lesser plasticity.

3.4 The *S* Equation

The psychological *situation* concept has been lamented by Rauthmann et al. [3] as being used often "haphazardly, ambiguously, [and] inconsistently" in the literature (p. 363). To ameliorate this, the authors proposed "three different basic kinds of situational information: *cues* (composition information), *characteristics* (psychological meaning information) and *classes* (category information)" (p. 363). Cues represent "physically present, scalable and (relatively) objectively quantifiable stimuli" (p. 364). Characteristics capture the "*psychologically important meanings* of perceived cues, thus summarizing a situation's psychological 'power'" (p. 364). Finally, classes represent "abstract groups, or types, of situations" (p. 364).

The formalization of Rauthmann et al.'s situation framework is now undertaken

$$S = F[Cu, Ch, Ce] \tag{8}$$

Where *S* equals "situation", *Ce* equals "class", *Ch* equals "characteristic", and *Cu* equals "cue". Given (3), (4), and (8) and the transitive property, (9) results

$$[P, S] = [F[Sl, Ot], F[Cu, Ch, Ce]] \tag{9}$$

By (9), person-situation interaction is extended as a function of the interaction between selves and others with cues, characteristics, and classes.

The following equation fuses (2)–(8), above

$$[F[Sl, Ot], F[Cu, Ch, Ce]] = F\left[St_{(T,Se)}, Pc\right]_{[(A,C,Mv,B)_{(U-,Sb-)Cs}]} \tag{10}$$

(10) reads: *Person-situation interaction as a function of self and other–cue, characteristic, and class interaction is a function of unconscious, subconscious, and conscious affective, cognitive, motivational, and behavioral traits, states, and processes.*

---

[12] How can Maslow's hierarchical view of motivation be compared with Kelly's more dynamic one? For Kelly, a more socially enactive and cognitive path would render motivation a clearer construct for psychological scientists (including personality psychologists) to operationalize.

4 Problem-Solving Utility of an Augmented Third-Force Framework

Solving massive socio-technical problems requires an equally massive scientific framework, like Sood's holarchic psychoinformatics. But what is this? Sood defined *psychoinformatics* as an interdisciplinary space joining psychology with computer science. Sood defined *holarchy* as a subjective-objective ontology with *holons* serving as basic units denoting part-wholes. Computer science is an objective domain, as is neurocognitive-behavioral psychology. Phenomenology, the study of human experience, is an ontologically *subjective* domain. Augmented cognition is a multi-ontological field depending on the mental paradigm employed by the researcher. Jackendoff's computational mind [20] is objective, but his phenomenological one is subjective. A holarchic approach to augmented cognition—or perhaps, more appropriately, augmented *mind*, if mind is granted to be broader than cognition—would mix both of these. From this view, and in keeping with this article's formalisms, augmented mind researchers would also need to consider augmented affect, behavior, and motivation in order to be holistically psychological (in a manner similar to [6]).

Augmented cognition updated to holarchic, mental augmentation would be simultaneously and equally a field interested in neurocognition and phenomenology (i.e., neurophenomenology in Varela's [21] proposed sense). Such a field could contribute to solving problems like climate change that are commonly known to be associated with cognitive deficits like confirmation bias (e.g., as demonstrated by skeptics) by generating and focusing on novel, positive concepts. An example of one such construct is "confirmation neutrality", which—rather than denoting the phenomenon of seeking evidence to confirm one's beliefs (as in confirmation bias)—denotes the phenomenon of seeking evidence to support only one's true beliefs (e.g.: "the sun will rise tomorrow"). The elaboration and study of such a construct would lend itself well to the epistemological augmented cognition researcher who defines knowledge in general as justified true belief.

Another construct that would be of great utility to the positive ecopsychologist in particular is seasonal affective "superorder" (SAS), as opposed to the informal clinical category seasonal affective disorder (SAD). While psychological disorder refers to fragmented person-situation interaction (i.e., mental and behavioral states and/or traits per (3)), such *superorder* would refer instead to integrated $[P, S] = F\big[St_{(T,Se)}, Pc\big]_{(M,B)}$. Theoretically necessary constructs for informal psychology that follow from SAD include "seasonal cognitive disorder" (SCD), "seasonal motivational disorder" (SMD), and "seasonal behavioral disorder" (SBD). For positive psychology, "seasonal cognitive superorder" (SCS), "seasonal motivational superorder" (SMS), and "seasonal behavioral superorder" (SBS) would follow from SAS.

"Superpersonality" and *super-mind* follow from Sect. 3.3's consideration of sub-personality and sub-mind, respectively. Taking these constructs further, one could also speak (however tentatively) of "superbehavior" (following from the existent *subbehavior*); *superconsciousness*; "subperson" and "superperson"; "subsituation" and "supersituation"; "subaffect" and "superaffect"; "supercognition" (particularly relevant for augmented cognition, and following from the existent *subcognition*); "submotivation" and "supermotivation"; and "subother" and "superother" (following from the existent *subself* and *superself*). The constructs encapsulated here by quotation marks remain to be elaborated philosophically, theoretically, scientifically, and practically.

"Cogfection" refers to affective cognition. *Cogfection* is relevant to the proposed System 3 type of thought, which is subconscious and both cognitive and affective [48]. In the psychological literature, intuition has generally been asserted as being affective and cognitive. However, recent discussions of System or Type [49] 3 thought vex this decision. If Type 3 is an average of System 1 and 2 processes—where System 1 consists of unconscious affect and System 2 consists of conscious cognition—then Type 3 thought stands as subconscious cogfect.

"Playing memory" (PM) should serve as a contrast construct to working memory. Working memory (WM) stores memory for more time than short-term memory (STM) does, but less than long-term memory (LTM). Thus, WM may be viewed as existing between STM and LTM in terms of storage duration; however, LTM's duration of storage (~90 s) is much closer to STM's (~30 s) than LTM's (potentially forever). Regardless, memory in general has received a sprawling treatment compared with its opposing (yet, cognitively- and neurologically-related—see [50]) construct *foresight*. Rather than storing previously encountered content like memory does, foresight projects future content. If WM serves as our support system for completing cognitive work, then PM should do the same with respect to affective play (though it might not serve such so much as accompany it freely). Adding foresight to the present discussion portends the additions of the following constructs: "short-term foresight" (STF), "working foresight" (WF), "long-term foresight" (LTF), and "playing foresight" (PF).

Lastly, the construct "wholicle" (pronounced whole-ical) is necessary for holarchic theory in general. Wholicles are to be contrasted with particles, and are intended to accompany the latter and follow from its role in the definition of holons as part-wholes [2]. In terms of current physics, wholicles refer to a level immediately above the macroscopic particle—i.e., the "microscopic wholicle" (all the way up to "macroscopic wholicle"). Macroscopic particles include powder, dust, stand, car parts, and galactic stars; microscopic wholicles refer to entities at least as large as galaxies. Such a term is more readily explicable in terms of current physics, but it is speculated that wholicles should refer more broadly and analogously to large entities within any holarchic domain. Within psychology, a wholicle might refer to the set of entities greater than persons or situations. Thus, the whole that is person-situation interaction represents the best starting point for such an advancement in holarchic-psychological theory.

5   Conclusion

Boden [32] explicated the evolution of theoretical psychology from the computational perspective (both cornerstones to the approach undertaken in this article and elsewhere [1, 2]) a la David Marr's work:

> "Marr argued that an adequate psychology will comprise explanations at three distinct but interrelated levels: computational, algorithmic, and hardware… This approach to theoretical psychology evolved through the late 1970s".

"The computational level...provides an abstract formulation of the information-processing task which defines a given psychological ability, together with a specification of the basic computational constraints involved. Where vision is concerned, these constraints are grounded in the structure of the physical world...they provide the necessary and sufficient computational basis for any creature (man, monkey, Martian—or machine) faced with the task in question".

"The second, algorithmic, level takes account of these constraints in specifying the psychological processes, or computations, by means of which the task is actually performed, which may differ in men and Martians. These processes are defined in terms of a particular system of representation, which can be proved to be reliable (to yield the relevant information) by reference to the top-level constraints" (pp. 49–50).

Marr's approach was similar to this article's in that both deem the informatic approach to theoretical psychology as necessary. The present article maintains skepticism regarding whether this approach is sufficient, though, particularly if phenomenological psychology is to be incorporated into a complete, holarchic psychology[13].

Boden also detailed Marr's methodology, which included hardware:

"The hardware-level deals with the neural mechanisms that embody the computational and algorithmic functions specified at the other two levels, showing how their psychological properties and anatomical connectivities enable them to do so. Hardware-properties may vary between species even more than algorithmic ones do, and are very different in machines".

"Marr's methodology was centered on computational, top-level, understanding of the nature of the information-processing task being modelled. He insisted that only if psychology is grounded in such understanding can be it a systematic science, as opposed to a ragbag of empirical findings, theoretically unjustified hunches, and *ad hoc* assumptions introduced to compensate for inadequacies in so-called theory" (p. 51).

Marr's informational-theoretic psychology thus included three levels: "computational", algorithmic, and hardware. It may be viewed as a predecessor to modern psychoinformatics, which has been discussed in numerous works [e.g., 1, 2]. A more complete psychology will someday successfully integrate phenomenological and computational psychologies. As Smith and Hamid [34] claim: "substantial improvement in AI could be achieved by adopting a hybrid framework in which embodied cognition...may contain representational abstract, and symbolic aspects" (p. 67). As these authors note, the potential for human-machine coexistence is strong.

---

[13] For the most famous and sprawling challenge to a purely computational approach to psychology, see [33].

References

1. Sood, S.: The psychoinformatic complexity of humanness and person-situation interaction. In: Arai, K., Bhatia, R. (eds.) FICC 2019. LNNS, vol. 69, pp. 496–504. Springer, Cham (2020). https://doi.org/10.1007/978-3-030-12388-8_35
2. Sood, S., et al.: Holarchic psychoinformatics: a mathematical ontology for general and psychological realities. In: Schmorrow, D.D., Fidopiastis, C.M. (eds.) HCII 2019. LNCS (LNAI), vol. 11580, pp. 345–355. Springer, Cham (2019). https://doi.org/10.1007/978-3-030-22419-6_24
3. Rauthmann, J., Sherman, R., Funder, D.: Principles of situation research: Towards a better understanding of psychological situations. Eur. J. Pers. **29**, 363–381 (2015)
4. Zahavi, D.: Self and Other: Exploring Subjectivity, Empathy, and Shame. Oxford University Press, Oxford (2014)
5. Buber, M.: I and Thou (1958)
6. Henriques G.: A New Unified Theory of Psychology. Springer, New York (2011). https://doi.org/10.1007/978-1-4614-0058-5
7. Klein, S.: The Two Selves: Their Metaphysical Commitments and Functional Independence. Oxford University Press, Oxford (2014)
8. Searle, J.: The Mystery of Consciousness. New York Review Books, New York (1997)
9. Varela, F., Shear, J.: The View from Within: First-Person Approaches to the Study of Consciousness. Imprint Academic, Thorverton (1999)
10. Cornelius, R.: The Science of Emotion: Research and Tradition in the Psychology of Emotion. Prentice-Hall, Upper Saddle River (1996)
11. Varela, F., Thompson, E., Rosch, E.: The Embodied Mind: Cognitive Science and Human Experience. The MIT Press, Cambridge (2000)
12. Lewin, K.: Field theory and experiment in social psychology: Concept and methods. Am. J. Soc. **44**, 868–896 (1939)
13. Skinner, B.F.: Generic nature of the concepts of stimulus and response. J. Gen. Psychol. **12**, 1240–1265 (1935)
14. Solipsism and the Problem of Other Minds. https://www.iep.utm.edu/solipsis/. Accessed 12 Nov 2019
15. Hovhannisyan, G., Henson, A., Sood, S.: Enacting virtual reality: the philosophy and cognitive science of optimal virtual experience. In: Schmorrow, D.D., Fidopiastis, C.M. (eds.) HCII 2019. LNCS (LNAI), vol. 11580, pp. 225–255. Springer, Cham (2019). https://doi.org/10.1007/978-3-030-22419-6_17
16. Hancock, M., Stiers, J., Higgins, T., Swarr, F., Shrider, M., Sood, S.: A hierarchical characterization of knowledge for cognition. In: Schmorrow, D.D., Fidopiastis, C.M. (eds.) HCII 2019. LNCS (LNAI), vol. 11580, pp. 58–73. Springer, Cham (2019). https://doi.org/10.1007/978-3-030-22419-6_5
17. Dennett, D.C.: The self as a center of narrative gravity. In: Kessel, F., Cole, P., Johnson, D. (eds.) Self and Consciousness: Multiple Perspectives. Erlbaum, Hillsdale (1992)
18. Freud, S.: The Interpretation of Dreams. (J. Strachey, Trans.) Basic Books, New York (1955). (Original Work Published in 1899)
19. Revelle, W.: Integrating personality, cognition, and emotion: putting the dots together? (2011). https://www.personality-project.org/revelle/publications/BPSP-revelle.pdf. Accessed 9 Jan 2019
20. Jackendoff, R.: Consciousness and the Computational Mind. The MIT Press, Cambridge (1987)
21. Varela, F.: Neurophenomenology as a methodological remedy to the hard problem. J. Consciousness Stud. **3**(4), 330–349 (1996)

22. McCrae, R., Costa, P.: Validation of the five-factor model of personality across instruments and observers. J. Pers. Soc. Psychol. **51**(1), 81–90 (1987)
23. Pervin, L.: Personality. In: Kazdin, A. (ed.) Encyclopedia of Psychology, vol. 6, pp. 100–106. American Psychological Association, Washington, DC (2000)
24. Shouse, E.: Feeling, emotion, affect. M/C J. **8**(6) (2005). http://journal.media-culture.org.au/0512/03-shouse.php. Accessed 18 Nov 2019
25. Kahneman, D.: Thinking, Fast and Slow. Farrar, Straus and Giroux [Kindle DX version] (2011)
26. Gore, J., Sadler-Smith, E.: Unpacking intuition: a process and outcome framework. Rev. Gen. Psychol. **15**(4), 304–316 (2011)
27. Maslow, A.H.: A theory of human motivation. Psychol. Rev. **50**, 370–396 (1943)
28. Kelly, G.: Clinical Psychology and Personality: The Selected Papers of George Kelly. John Wiley & Sons Inc., New York (1969)
29. Dweck, C., Leggett, E.: A social-cognitive approach to motivation and personality. Psychol. Rev. **95**(2), 256–273 (1988)
30. Proctor, C., Tweed, R., Morris, D.: The Rogerian fully functioning person: a positive psychology perspective. J. Humanist. Psychol. **56**(5), 1–28 (2015)
31. Horley, J.: The units of analysis problem in psychology: an examination and proposed reconciliation. In: Baker, W.J., Mos, L.P., Rappard, H.V., Stam, H.J. (eds.) Recent Trends in Theoretical Psychology. Recent Research in Psychology. Springer, New York (1989). https://doi.org/10.1007/978-1-4613-9688-8
32. Boden, M.A.: Computer Models of Mind. Cambridge University Press, Cambridge (1984)
33. Dreyfus, H.L.: What Computers Still Can't Do: A Critique of Artificial Reason. [Kindle DX version] (1999)
34. Smith, N.L., Hamid, O.H.: Embodied cognition and human-machine coexistence. UKH J. Sci. Eng. **1**(1), 67–71 (2017)
35. DeYoung, C.G., Weisberg, Y.J., Quilty, L.C., Peterson, J.B.: Unifying the aspects of the Big Five, the interpersonal circumplex, and trait affiliation. J. Pers. **81**(5), 465–475 (2013)
36. Horney, K.: Neurosis and Human Growth: The Struggle Toward Self-Realization. W. W. Norton & Company, New York (1991). (Original Work Published in 1950)
37. Plato: The Republic. (2nd ed.). Penguin Books, London (2007). (Original Work Published in ~ 375 B.C.)
38. Maslow, A.H.: Toward a Psychology of Being. (1st ed.). Wilder Publications, Blacksburg (2011). (Original Work Published in 1962)
39. Hofstadter, D.: Gödel, Escher, Bach: An Eternal Golden Braid (20th Anniversary ed.). Basic Books, New York (1999)
40. Klein, S.: The self and science: Is it time for a new approach to the study of human experience? Curr. Dir. Psychol. Sci. **21**(4), 253–257 (2012)
41. Maslow, A.H.: Motivation and Personality. Harper & Row Publishers, New York (1954)
42. Gerbasi, M.E., Prentice, D.A.: The self- and other-interest inventory. J. Pers. Soc. Psychol. **105**(3), 495–514 (2013)
43. Rabani, E., Perg, L.: Demonstrably safe self-replicating manufacturing systems: banishing the halting problem—organizational and finite state machine control paradigms. In: Schmorrow, D., Fidopiastis, C. (eds.) Augmented Cognition. HCII 2019. Lecture Notes in Computer Science, vol 11580. Springer, Cham (2019)
44. McGonigal, J.: SuperBetter: A Revolutionary Approach to Getting Stronger, Happier, Braver, and More Resilient. Penguin Press, New York (2015)
45. Yalom, I.D.: The Gift of Therapy: An Open Letter to a New Generation of Therapists and Their Patients. Harper Perennial, New York (2002)
46. DARK STAR (1974). http://www.cinemah.com/altri/war/carpent.htm. Accessed 13 Jan 2020

47. Kukla, A.: Personality Theory: A Book of Readings. Canadian Scholars' Press, Toronto (1996)
48. Slingerland, E.: Trying Not to Try: The Art and Science of Spontaneity. Crown Publishers, New York (2014)
49. Dijksterhuis, A., Strick, M.: A case for thinking without consciousness. Per. Psychol. Sci. **11**(1), 117–132 (2015). https://www.researchgate.net/publication/292188707_A_Case_for_Thinking_Without_Consciousness. Accessed 15 Jan 2020
50. Suddendorf, T.: Episodic memory versus episodic foresight: similarities and differences. WIREs Cog. Sci. **1**, 99–107 (2010)
51. Bandura, A., Huston, A.C.: Identification as a process of incidental learning. J. Abnorm. Soc. Psych. **63**(2), 311–318 (1961)

A Preliminary Experiment on the Evaluation of Aroma Effects Using Biological Signals

Runqing Zhang[✉], Chen Feng, Peeraya Sripian, and Midori Sugaya

Shibaura Institute of Technology, Tokyo, Japan
{ma19053,peeraya,doly}@shibaura-it.ac.jp

Abstract. Aromatherapy is a natural way of healing a person's mind and body. The essential oils involved in aromatherapy are highly concentrated essences extracted from the plants through the process of the distillation. Each oil produces a predictable and reproducible effect on the user when its aroma is inhaled. There have been lots of researches that confirm the effects of inhaling aroma by using biological signals. However, most of these researches, only the heart rate or the pulse signals are used. And, electroencephalograms (EEGs) is rarely used in evaluating the aroma effects. So in order to increase the validity of the reputed effects of the previous research, we consider to use the combination of the EEG signal and the pulse signal to evaluate the effects of the aroma. This work is the preliminary experiment to evaluate the changes in this combination of biological signals after inhaling several aromas. In this work, we found that the aromas can have different effects at the same time in the EEGs and the pulses.

Keywords: Aroma · Heart rate · EEG

1 Introduction

In today's high-paced, efficient modern society, stress has become an indispensable part of daily life. Self-regulation skills can help us cope with demanding situations, reduce negative health outcomes, and increase overall quality of life [1]. Learning how to self-regulate is an important skill because of the stressors can't be reduced. There are various ways for people to regulate the stress by themselves, including yoga and meditation. While these methods can be effective in helping manage stress, they have some disadvantages. For example, Self-guided interventions, including meditation and yoga, suffer from high dropout rates [2] due to the unengaging nature of the exercises and lack of motivation [3]. In addition, these methods require a specific environment, somewhere quiet and open.

Aromatherapy is well suited to address the shortcomings of existing methods for stress management. Aromatherapy derived its name from the word aroma, which means fragrance or smell and therapy which means treatment [4]. This therapy method is a natural way of healing a person's mind or body by using the essential oils, as the main therapeutic agents. The essential oils have found their essence as aromas with a curative potential on the body and mind [5], and different aromas have different effects. There are

© Springer Nature Switzerland AG 2020
D. D. Schmorrow and C. M. Fidopiastis (Eds.): HCII 2020, LNAI 12196, pp. 94–104, 2020.
https://doi.org/10.1007/978-3-030-50353-6_7

many types of methods by which they are controlled in small quantity like inhalation, local application, and baths. Today the popularity of aromas for pleasure, relaxation, and therapeutics is unabated and typified in the ever-popular application of aromatherapy.

In the stress analysis using the autonomic nervous system, methods for evaluating various physiological responses such as heartbeat, respiration, and pulse wave in addition to vital signs such as EEG, facial temperature, skin surface temperature and surface potential, and eye movements are being studied. The advantage of these analysis methods is that they can be measured non-invasively, and biological information can be measured by configuring amplifiers and filters, so they can be implemented at a relatively low cost. Aromas are found can effectively reduce the level of stress [6].

Diego et al. found the EEG readings to show increased beta-power following lavender inhalation [7], however Duan et al. found that after the lavender aroma stimulus as seen as increases in the HF component and decreases in the LF/HF [8]. So in order to increase the validity of the reputed effects of the previous research, we consider to use the combination of the EEG and the pulse signal to evaluate the effect of the aromas.

In this study, we evaluate the effect by this method as a preliminary study. We focused on how the EEG and the pulse signal will change to know the effect of the aroma stimulus. Also to know the differences between the subjective effect and objectively analyzed effect by the sensors, we use SAM questionnaire that can evaluate the subjective emotion with the arousal and valence attributions. We compare the results from sensors and the data. We conducted the experiment with 3 collaborators. They inhaled the aromas of essential oils with sensors and obtain the data to analyze the detailed effect of the aroma. Then, the result show that there is no significant correlation between the EEG and the results of subjective evaluation.

This paper is organized as follows. Section 2 reviews prior research on the aromatherapy which using biological signals. Section 3 describes the method we proposed in this work. Section 4 describes the preliminary experimental protocol. And Sect. 5 describes the analysis of the experimental results from the study. The article also concludes with a thorough discussion of these results and directions for future work.

2   Literature Review

Aroma have been used for their medicinal and mood altering properties throughout history, and aroma molecules have direct effects on human behavior and physiology ranging from activation of memories to changes in mood or emotional states [7]. Each oil produces a predictable and reproducible effect on the user when its fragrance is inhaled [9]. There have been lots of researches that confirms the effects of inhaling essential oils. For example, Warm et al. evaluated that the peppermint fragrance could enhance attention, and improve performance on visual vigilance tasks; meanwhile, the lavender fragrance could decrease anxiety and tension [10]. Employing more participant dependent measures, Diego et al. found the electroencephalograms (EEGs) readings to show increased beta-power following lavender inhalation, implying neurological sedation and corroborating subjective reports of calmness, whereas jasmine has been demonstrated to produce increased alpha-power in the frontal cortices, indicative of increased arousal, and they also described the beneficial effect of the lavender and the orange fragrance on the measure of implicit memory [7, 11]. On the other hand, Han et al. found that the aromas have effect on the autonomic nervous activity and suggested that the aromas are effective in stress management [12].

In those works, the effects of the aromas were validated mainly by comparing one of the biological signals like EEGs taken from the participant's before and after the aromas were inhaled. Mostly the signals that were used, are the heart rate, or the pulse signals. Furthermore, they also used the subjective questionnaire in order to know how the participant is feeling.

However, the brainwaves can identify the changes in the brain activity, and the pulse can identify the changes on the autonomic nervous activity. And the participants may not answer how they are exactly feeling. In order to increase the validity of the reputed effects of the subjective questionnaire, we want to find how the brain activity and the autonomic nervous activity will be changed at the same time. We propose a method that combines the EEGs and the pulse signals to make a more comprehensive evaluate the effects of aromas.

3   Proposed Method

To solve the existent study issue, we consider to use a combination of the EEG and the pulse signal to evaluate the effect of aroma. The Fig. 1. shows the proposed system. Based on previous research, we support the combination between the measurement of the EEG signal and the measurement of the pulse signal. In this way, we record the EEG and the pulse signals of participants during exposure to the aroma stimulation. The EEG' power can show the level of arousal about the brain activity, and the pulse signals can show the level of valence about the autonomic nervous activity. We want to find out how the participant's biological signals would change after inhaling the essential oil aroma and the relationship between the questionnaire and the biological signals.

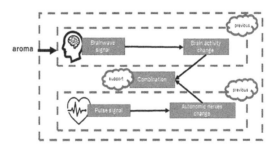

Fig. 1. Overview of the system

3.1 Brainwave

To know the effect of the brain, we consider the point of the brain signals. The largest portion of the human brain, the cortex, is divided into the frontal, temporal, parietal, and occipital lobes [13]. The frontal lobe is responsible for the conscious thought, and according to the International 10/20 System (IS), the AF3 electrode on the frontal lobe is the easiest way to measure.

These potential differences as signals are divided into specific ranges that are more prominent in certain states of mind, namely the δ (1–4 Hz), δ (4–7 Hz), α (8–13 Hz), β (13–30 Hz), and γ (>30 Hz) bands [13].

δ brainwaves are associated with the generated in deepest meditation and dreamless sleep. θ brainwaves occur most often in sleep however are also dominant in deep meditation. α brainwaves are typically associated to a relaxed mental state, and aid overall mental coordination, calmness, alertness, mind/body integration and learning. The α brainwaves are further divided in tow bands. Low-α (α1, 8–10 Hz) and High-α (α2, 10–13 Hz). The β waves dominate normal waking state of consciousness when we are alert, attentive, engaged in problem solving, judgment, decision making, or focused mental activity. The β brainwaves are also divided in tow bands. Low-β (β1, 13–22 Hz) and High-β (β2, 22–30 Hz). Finally, γ waves are associated with a hyper brain activity [14].

In this work, we use the values about Low-Alpha, and Low-Beta to evaluate the level of arousal. This value can be obtained using the following equation;

$$x = \begin{cases} \frac{High\beta}{High\alpha}, & highx \\ \frac{Low\beta}{Low\alpha}, & lowx \end{cases} \tag{1}$$

The lower value indicates the lower level of brain activity (Sleepiness), and the higher value indicates the higher level of brain activity (Arousal). To evaluate the brain wave, we use NeuroSky's MindWave Mobile to measure the brain signals in this work. The NeuroSky's MindWave Mobile records the potential difference between the ear lobe and the frontal lobe (AF3), and collect the data with proposed indexes.

3.2 Pulse Signals

The heartbeat interval is the time between the beat of the heart beat and the next beat [15]. Therefore, when a human is lying down or resting, the pulsation interval increases, so the heart beat interval becomes long. Also, when a human performed exercise or obtained nervous, the heart rate becomes slow, based on the idea, the heartbeat interval decreases. The lower the value, the more times the heart is beating. The lower the value is, the greater the number of times the heart is beating. We use this value to evaluate the participant's stress in this study. The pNN50 refers to a ratio of consecutive heartbeat interval differences of 50 ms or more, and it is thought that the larger this value is, the more the "comfort state" becomes [15].

In this research, we used Switch Science's Pulse Sensor [16] as the pulse sensor. This pulse sensor is one of Arduino's sensors, and by wearing it on an earlobe or a fingertip, pulse information of the wearer can be acquired. The pulse information is sent to the program in the computer via Arduino. With this pulse sensor, various values related to heartbeat can be acquired. The value used in this research is described below.

4   Preliminary Experiment

4.1   Experiment Contents

We conducted a preliminary experiment. The collaborator of the experiment is one female in 20's.

Following the proposed method, we set up the experimental procedure based on the experimental protocol [7]. The detailed procedures are shown in the Fig. 2.

In this experiment, the participants are required to wear an EEG and a pulse sensor, answer a questionnaire about personal information, and to close their eyes in order to stay relaxed during the whole trial before the beginning of the experiment. After two minutes, the olfactory of the participant is stimulated by a test strip with essential oil aroma for one minute. Then, the participants are required to answer another questionnaire (SAM) about how their feeling during inhaling these aromas. Furthermore, the participants must clean their hands before the experiment, because between the two aromas, the olfactory of the participant need to be reset by smelling the skin at wrist [17], and have a ten minutes' rest.

The essential oils we prepared are the 1. Chamomile, 2. Lavender, 3. Orange, 4. Peppermint. For these four types of essential oils, each of the concentration was prepared in two concentrations, 10% and 50%, and compared with those at rest. These aromas are used as the olfactory stimulation.

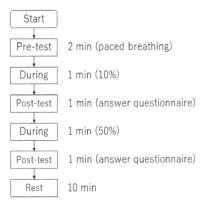

Fig. 2. Experiment protocol. 10%: 10% concentration, 50%: 50% concentration

4.2 Subjective Questionnaire

In order to consider about the participants' self-understanding of the aroma, we use the questionnaire named SAM.

The SAM (Self-Assessment Manikin), is a non-verbal pictorial assessment technique that directly measures the pleasure, arousal, and dominance associated with a person's affective reaction to a wide variety of stimuli. In this experiment, 1 representing is the most arousal and the most pleasure, 9 representing the most sleepiness and the most misery. The SAM is a psychological questionnaire that can track the personal response to the affective stimulus, and it is an unexpansive, effective method for quickly assessing reports of affective response in many contexts [18].

4.3 The Analysis of the Result

Figure 3 shows that the participant was changed after inhaled the Chamomile aroma. We can see that the level of Low-x was not be changed, however the level of High-x was changed ($p < 0.05$). Both of the x is higher than 1. High-x is higher than 1 that indicates that the arousal level of the brain was higher, and the participant's brain became more active than before inhaling this aroma. However, the pNN50 below 0.3 during the 10% concentration, higher than 0.3 during the 50% concentration. The value of the pNN50 is higher than 0.3 that indicates that the stress was decreased at the 50% concentration, and the participant felt more pleasure.

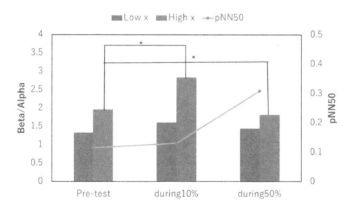

Fig. 3. Biological signals result of the before and after Chamomile aroma inhalation

Figure 4 shows that the participant was changed after inhaled the Lavender aroma. We can see that during 10% concentration, the level of Low-x was decreased and below 1 ($p < 0.05$). This indicates that the sleepiness level of the brain was higher, and the participant's brain became more sleepiness. While the level of High-x was not changed obviously. The pNN50 ($p < 0.05$) shows that the stress was not decreased. This result was the same as the subjective questionnaire.

During 50% concentration, the result is different from the subjective questionnaire. The figure shows that the participant's brain became active and not much pleasure. However, the participant felt more arousal and pleasure.

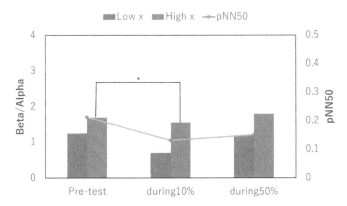

Fig. 4. Biological signals result of the before and after Lavender aroma inhalation

Figure 5 shows that the participant was changed after inhaled the Orange aroma. We can see that during 10% concentration, the level of Low-x was increased and higher than 1 (p < 0.05). This indicates that the arousal level of the brain was higher, and the participant's brain became more arousal. While the level of High-x was not changed obviously. The pNN50 (p < 0.05) shows that the stress was increase. However it is different from the subjective questionnaire, the participant felt too much pleasure in this concentration.

During 50% concentration, the figure shows that the participant's brain was not too much arousal, and a little pleasure. This indicates that the stress was decreased after inhaling the 50% concentration of the orange aroma.

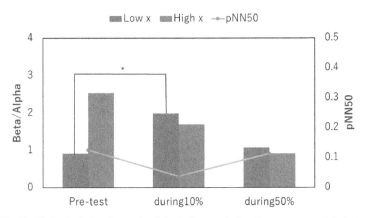

Fig. 5. Biological signals result of the before and after Orange aroma inhalation

Figure 6 shows that the participant was changed after inhaled the Peppermint aroma. We can see that during 10% concentration, the level of Low-x was increased and higher than 1 (p < 0.05). This indicates that the arousal level of the brain was higher, and the participant's brain became more arousal. While the level of High-x was not changed obviously. The pNN50 shows that the stress was not decreased too much. However it is different from the subjective questionnaire, the participant felt too much sleepiness and pleasure in this concentration.

During 50% concentration, the figure shows that the level of Low-x was increased and higher than 1 ($p < 0.05$), the participant's brain was not too much arousal. The pNN50 ($p < 0.01$) shows that the stress was decreased obviously. This indicates that after inhaling the Peppermint aroma, no matter which concentration, the stress can be decreased.

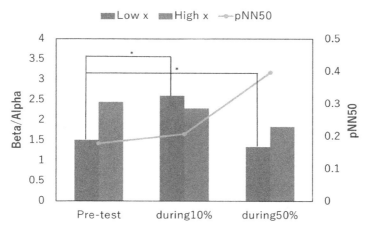

Fig. 6. Biological signals result of the before and after Peppermint aroma inhalation

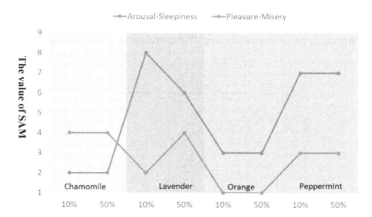

Fig. 7. The subjective questionnaire results after inhaled the aromas

We executed the preliminary experience for the all of the experiment's participant. The Fig. 7 shows that after inhaling the aroma, the subjective evaluation of the aroma by the participant. We found that the two concentrations of the aromas were the almost same for the participant, except for lavender. And in the 10% concentration of the lavender aroma, the participant showed 8 level of the sleepiness and 2 level of the pleasure. Something other phrases, we consider the 10% concentration of lavender aroma has the most significant stress relief effect, and it also has the effect of relaxing and helping sleep.

4.4 Relation Analysis

We also analyzed the relationship between the result of the subjective questionnaire and the result of the biological signals. Since we set the most pleasure as level 1, the most misery as level 9 in the subjective questionnaire, the negatively related result support to show the positive related between the result of the subjective questionnaire and the result of the biological signals. After regression analysis of the EEG results and the subjective evaluation results, as shown in Fig. 8, we found that there is no significant correlation between the EEG and the results of subjective evaluation.

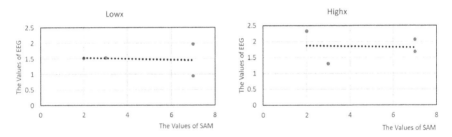

Fig. 8. The regression analysis about the EEG signals

However, in the regression analysis of the pulse results and the results of the subjective questionnaire results, we found a clear negative correlation between these two, as shown in Fig. 9.

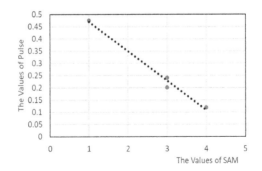

Fig. 9. The regression analysis about the pulse signals

As shown in Table 1, although the average of the EEG signals data has changed, the P value is not obvious. And The correlation between EEG signals data and subjective evaluation is low. The approximate curve close to a straight line. In this experiment, we did not describe the name of this aromas to the participant. Therefore, we consider that in this case, the subjective evaluation of the aroma is not related to the cognition of the participant. However, there is a high correlation and an obvious P value between the pulse signal and the subjective evaluation. The approximate curve presents one-dimensional linear correlation. Therefore, we consider in this case, the subjective evaluation of the aroma correlates with changes in the result of the pulse signals. In other words, we can use the changes of the pulse signals to decide whether the participant is pleasure or misery after inhaling the aroma. Meanwhile, the aromas used as stimuli in this experiment was selected with the concentration were in the low and medium range. We assumed that there is a possibility that the perception of aroma concentration in our body is different from the perception of aroma concentration in our cognition.

Table 1. The analysis about the relationship between the subjective questionnaire and the biological signals (*$p < 0.05$, **$p < 0.01$)

	Correlation	p-Value
Low-x	-0.092	0. 095
High-x	-0.075	0. 120
pNN50	-0.993	0. 036*

5   Conclusion

We proposed the combination method from the biological signals and aroma and compared with subjective evaluated the effect of aromas in this research. We did the experiment about the aromas were inhaled by the participant and the participant's biological signals were recorded. And in these different aromas, we found that it can have different effects on the EEG and the pulse at the same time. Although there is no significant difference observed because of the limited number of participants, it can be implied that aroma can become a Self-regulation way to decrease the stress. And in this work, by analyzing subjective evaluation and bio signals, the results of subjective evaluation have a clear relationship with the pulse signals.

5.1   Future Work

From the result, we found that the combination of biological signals can be used to evaluate the effect of the aromas. In the future, we want to change the condition that before the experiment, let the participant to know the name of the aroma. And more participants are necessary for more reliability of the data analysis. Experiment procedure could be improved such as selection of stimuli, presented time, record time and so on. And also improve the concentration of the essential oils that could induce more significant differences before and after inhaling these aromas. Furthermore, we want to classification these aromas by the Circumflex Model of Affect [19]. It could be possible to find out if the participant's preferences affect the effect of essential oil aroma by analyzing the biological signals.

References

1. Parnandi, A., Gutierrez-Osuna, R.: Visual biofeedback and game adaptation in relaxation skill transfer. IEEE Trans. Affect. Comput. **10**(2), 276–289 (2017)
2. Rose, R.D., et al.: A randomized controlled trial of a self-guided, multimedia, stress management and resilience training program. Behav. Res. Ther. **51**(2), 106–112 (2013)
3. Davis, M.J., Addis, M.E.: Predictors of attrition from behavioral medicine treatments. Ann. Behav. Med. **21**(4), 339–349 (1999)
4. Ali, B., Al-Wabel, N.A., Shams, S., Ahamad, A., Khan, S.A., Anwar, F.: Essential oils used in aromatherapy: a systemic review. Asian Pacific J. Trop. Biomed. **5**(8), 601–611 (2015)
5. Moss, M., Hewitt, S., Moss, L., Wesnes, K.: Modulation of cognitive performance and mood by aromas of peppermint and ylang-ylang. Int. J. Neurosci. **118**(1), 59–77 (2008)
6. Seo, J.Y.: The effects of aromatherapy on stress and stress responses in adolescents. J. Korean Acad. Nurs. **39**(3), 357–365 (2009)
7. Diego, M.A., et al.: Aromatherapy positively affects mood, EEG patterns of alertness and math computations. Int. J. Neurosci. **96**(3–4), 217–224 (1998)
8. Duan, X., et al.: Autonomic nervous function and localization of cerebral activity during lavender aromatic immersion. Technol. Health Care **15**(2), 69–78 (2007)
9. Sanderson, H., Ruddle, J.: Aromatherapy and occupational therapy. Br. J. Occupat. Ther. **55**(8), 310–314 (1992)
10. Warm, J.S., Dember, W.N., Parasuraman, R.: Effects of olfactory stimulation on performance and stress. J. Soc. Cosmet. Chem. **42**(3), 199–210 (1991)
11. Degel, J., Piper, D., Köster, E.P.: Implicit learning and implicit memory for odors: the influence of odor identification and retention time. Chem. Sens. **26**(3), 267–280 (2001)
12. 韓在都, & 内山明彦. 嗅覚刺激が生体に与える影響の計測と解析. 電気学会論文誌. C, **122**(9), 1616–1623 (2002)
13. Niedermeyer, E., da Silva, F.L. (eds.): Electroencephalography: Basic Principles, Clinical Applications, and Related Fields. Lippincott Williams & Wilkins, London (2005)
14. Bos, D.O.: EEG-based emotion recognition. Influence Visual Auditory Stimuli **56**(3), 1–17 (2006)
15. Hjortskov, N., Rissén, D., Blangsted, A.K., Fallentin, N., Lundberg, U., Søgaard, K.: The effect of mental stress on heart rate variability and blood pressure during computer work. Eur. J. Appl. Physiol. **92**(1–2), 84–89 (2004)
16. Switch Science, 心拍センサ (2010). https://www.switch-science.com
17. Grosofsky, A., Haupert, M.L., Versteeg, S.W.: An exploratory investigation of coffee and lemon scents and odor identification. Percept. Mot. Skills **112**(2), 536–538 (2011)
18. Bradley, M.M., Lang, P.J.: Measuring emotion: the self-assessment manikin and the semantic differential. J. Behav. Ther. Exp. Psychiatry **25**(1), 49–59 (1994)
19. Posner, J., Russell, J.A., Peterson, B.S.: The circumplex model of affect: An integrative approach to affective neuroscience, cognitive development, and psychopathology. Dev. Psychopathol. **17**(3), 715–734 (2005)

Electroencephalography and BCI

A Neuroimaging Approach to Evaluate Choices and Compare Performance of Tower Air Traffic Controllers During Missed Approaches

Alvin J. Ayeni[1(✉)], Kiranraj Pushparaj[1], Kurtulus Izzetoglu[2], Sameer Alam[1], and Vu N. Duong[1]

[1] Nanyang Technological University, Air Traffic Management Research Institute,
65 Nanyang Drive, Singapore 637460, Singapore
`alvin.ayeni@ntu.edu.sg`
[2] Drexel, University School of Biomedical Engineering, Science and Health Systems,
3141 Chestnut Street, Philadelphia, PA 19104, USA

Abstract. The aim of this research is to use functional Near-Infrared Spectroscopy to compare and contrast brain activation for professional versus novice Tower Air Traffic Controllers when performing their daily tasks, whilst accounting for missed approaches. With functional Near-Infrared Spectroscopy chosen due to its ability to continuously monitor brain activity for mobile participants in their workplace settings, increasing ecological validity, as well as being safe, inexpensive, and benefitting from low set-up times, resulting in excellent temporal resolution as well as superior spatial resolution over Electroencephalogram. If a significant difference in activation is observed between professional and novice ATCOs, the neuroimaging data can be used as a benchmark for future exploratory studies using the obtained neuroimaging data to serve as a reliable quantitative measure to track performance during Air Traffic Controller training, establishing a metric to distinguish novice from professional Air Traffic Controllers. Our hypothesis is that professional tower controllers will have a decrease in brain activation due to their experience. Contrastingly, novice tower controllers would have more extensive brain activation, given a lack of experience relying soley on training. Additionally, we expect to see a significant difference in sustained attention activation between professionals and novices. The tasks that the tower controllers will be expected to resolve will be a series of tower control duties that will be severely impacted by a range of factors that will intentionally make the successful performance of their duties strained.

Keywords: Human factors · Neuroscience · Neuroergonomics · Air traffic management · Neuroimaging · Brain-Computer interaction

1 Introduction

Human factors remains an integral part of Air Traffic Management (ATM) (Rodgers 2017; Leveson 2001). Despite technological advances in ATM systems and operations,

© Springer Nature Switzerland AG 2020
D. D. Schmorrow and C. M. Fidopiastis (Eds.): HCII 2020, LNAI 12196, pp. 107–117, 2020.
https://doi.org/10.1007/978-3-030-50353-6_8

the human operators of these systems continue to be a crucial aspect in maintaining the safety and efficiency for successful task completion (Billings 2018; Rodgers 2017). Indeed, past ATM research has shown that the decision aiding systems currently at the disposal of Air Traffic Controllers (ATCOs) are at times disregarded or not used in the manner that they were designed, leading to misuse and disuse (Westin et al. 2015; Bekier et al. 2015). Hence, it is important that cognitive workload and performance remain at optimal levels to ensure a productive and safe working environment. Neuroergonomics, the amalgamation of neuroscience and human factors provides insight into these human factors by applying neuroscience to real-world working environments to address topics such as, but not limited to, individual differences in cognition, human error, operational safety, cognitive workload, and performance by utilising various neuroimaging devices and stimulation techniques (Arico et al. 2017; Parasuraman 2011; Fedota and Parasuraman 2010). Using neurophysiological measures as evaluators for optimal workload, and peak performance instead of previous subjective reports that do not consider the brain's current state (Curtin and Ayaz 2018). Allowing for greater accuracy and reliability in assessing human performance in naturalistic working environments, which will be especially pertinent into how new technologies and paradigms are developed in the future (Arico et al. 2017; Curtin and Ayaz 2018; Izzetoglu and Richards 2019; Derosière et al. 2013).

As air travel increases, current capacities will be exceeded within a decade (ICAO 2014). Based on projections such as these alongside airport expansion constraints, ATCOs are estimated not to be able to handle traffic that is 25% above present levels (FAA 2015). Leading to a degradation of performance by ATCOs and therefore, an increased chance of incidents occurring (Kirwan 2001). Furthermore, with human errors accounting for most operational errors that occur in the ATM environment, it is of utmost importance that these issues are minimised or in best case scenarios, eliminated entirely (Redding 1992). Being particularly crucial for Tower Controllers, considering Boeing found that 44% of all fatalities and 49% of fatal accidents globally, from 2008–2017, occurred during final approach and landing (Boeing Commercial Planes 2018). One of the tasks of Tower Controllers, that is infrequent, but can drastically affect performance and productivity, is a missed approach procedure leading to a go-around, where an aircraft rejects the chance to land during final approach and makes a subsequent attempt (Dai et al. 2019). Experience plays a key role in managing these complex situations, as it can help ease the demands associated with having to make quick and accurate decisions in a highly dynamic environment, by using a range of varied past experiences to help mediate new ones (Seamster et al. 1993; Liu and Wickens 1994). More specifically, tower controllers have the arduous responsibility in managing all of the movements made by aircrafts or vehicles within the vicinity of the airport, issuing taxi instructions, managing the clearances needed for take-off and landing, issuing Instrument Flight Rules (IFR), flight plan clearances, and transponder codes for Visual Flight Rules (VFR) aircraft. All whilst dealing with more traffic than ever before, that is only expected to increase (FAA 2015).

The aim of this research is to use a quantitative neuroimaging approach, specifically employing functional Near-Infrared Spectroscopy (fNIRS) to compare and contrast sustained attention and decision-making related brain activation for professionally-trained and unprofessionally-trained Tower ATCOs, when executing complex non-nominal events such as handling multiple arrivals, whilst accounting for missed approaches. fNIRS has been utilized here for its ability to continuously monitor brain activity for mobile participants in their workplace settings, increasing ecological validity. Moreover, fNIRS is safe, inexpensive, and benefits from low set-up times, resulting in excellent temporal resolution as well as superior spatial resolution over Electroencephalogram (EEG) allowing for more accuracy in identifying the brain regions associated with specific tasks (McKendrick et al. 2015). fNIRS is able to measure the differences in brain activation by focusing near-infrared light on the scalp, in order to measure changes in blood oxygenation as neurons that are more associated with specific tasks require more energy, or oxygenated haemoglobin (Izzetoglu, et al. 2004). fNIRS measures the cerebral hemodynamic response or the conversion of oxy-haemoglobin to deoxy-haemoglobin using specific wavelengths of light from the absorption spectrum regions of 650 to 950 nm, as the absorption spectra of oxy-haemoglobin and deoxy-haemoglobin are distinct in this region, allowing for the determination of concentration change between the two via scattered light diffusion measurement (Jobsis, 1977; Ferrari et al. 2004).

If a significant difference in activation is observed between professional and novice ATCOs, the neuroimaging data can be used as a benchmark for a future exploratory utilising neuroimaging to assist in neurostimulation via transcranial Direct Current Stimulation (tDCS). Enabling ATCOs to be better equipped to manage the rigors of their duties using neuroimaged-guided neurostimulation. This method could potentially mitigate cognitive deficiencies, or momentary lapses in judgement that ATCOs could potentially face due to stress and fatigue (Ayaz et al. 2012). Indeed, previous studies have suggested tDCS on specific brain regions can augment the performance of tasks associated with that region. (Fregni et al. 2005; Dockery et al. 2009; Coffman et al. 2014; Nelson et al. 2012; Parasuraman and McKinley 2014; Coffman et al. 2012; Falcone et al. 2012; Nelson et al. 2014; Clark et al. 2012). The proposed experimental paradigm would be to identify the executive decision-making brain regions that are most activated when operators are under high cognitive loads using the hemodynamic response obtained through fNIRS selected over other neuroimaging device due to an optimal level of both spatial resolution and portability (Parasuraman 2011). Though attentional networks are interconnected with various regions of the brain, the frontal regions have traditionally been most associated with tasks involving sustained attention or vigilance (Parasuraman et al. 1998). Past studies conducted using other neuroimaging devices with superior spatial resolution compared to fNIRS, ranging from Positron Emission Topography (PET) to functional Magnetic Resonance Imaging (fMRI) have implicated the lateral portions of the frontal lobe, for vigilance-related tasks; with particular activation of the right lateral frontal cortex (Parasuraman 2011; Berman and Weinberger 1990; Cohen et al. 1992; Coull et al. 1998; Lewin et al. 1996; Pardo et al. 1991). Moreover, it was discovered that patients who suffered from right frontal lesions, had increased reaction times and missed more targets in the Continuous Performance Test (CPT), even when their performance was contrasted with patients who had left and bilateral frontal lesions (Rueckert and Grafman 1996).

Suggesting that the longer vigilance is sustained, the more crucial the right frontal region becomes (Rueckert and Grafman 1996).

Additionally, any significant differences observed between professional and novice ATCOs, can be used for another potential study in which the neuroimaging data could serve as a reliable quantitative measure to track performance during training, thus establishing a metric to distinguish novice ATCOs from experts.

In this study, our hypothesis is that experienced tower controllers will implement more varied and innovative problem-solving strategies to resolve these situations which will be reflected by an decrease in associated brain region activation as their years of experience augments their ability to resolve conflicts faster. Contrastingly, novice tower controllers will not be at the liberty to use more varied problem solving skills and must rely solely on their training, resulting in more extensive brain activation (Liu and Wickens 1994; Milton et al. 2007). The tasks that the tower controllers will be expected to resolve will be a series of tower control duties that will be severely impacted by a range of factors that will intentionally make the successful performance of their duties strained, such as runway capacity limitations, high density of ground movement, and most cognitively taxing, re-routing missed approach flights (Fig. 1).

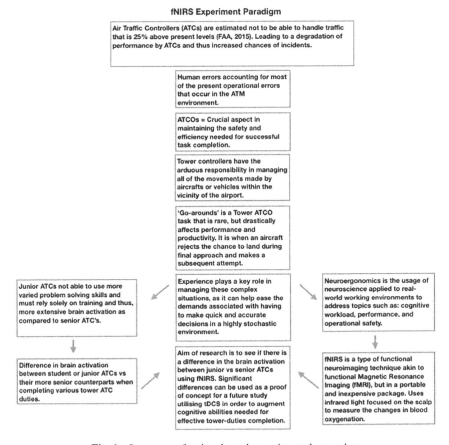

Fig. 1. Summary of rationale and experimental procedure

2 Materials and Methods

2.1 Ethics Statement

All participants provided informed consent to take part in this study, which was approved by the Nanyang Technological University Institutional Review Board

Participants. A total of six participants (all male) from the Air Traffic Management Research Institute (ATMRI) volunteered to participate in the study. Two of the participants were classified as novice ATCOs given that they had no professional training for tower ATCO duties. Another two were classified as professional ATCOs given that they had previous professional training in tower ATCO duties. The remaining 2 participants served as pseudo-pilots in the scenarios. The pseudo-pilots communicated to the ATCOs if an incoming flight would be a missed approach.

ATCO Vigilance Task. The vigilance task used in this experiment was a simulated tower ATCO task in which, participants had to actively manage 15 arrivals into Changi Airport, with 5 of those 15 scheduled flights being classified as a missed approach requiring the aircraft to execute a go-around procedure and attempt to land again at a future point. Artificially increasing the workload, whilst calling for participants to stay particularly vigilant in order to avoid potential collisions with other aircraft moving in and around the aerodrome. Moreover, the simulation included heavy rain and reduced visibility to mimic the real- world environments in which a missed approach would be most likely. 20 flights were chosen to mimic the average workload a tower controller will experience at Changi Airport based on data taken from 4:01 pm on the 15th of December 2019, to 4:01 pm 16th of December. The vigilance task took approximately 1 h for completion. Before the vigilance task commenced, the novice ATCOs partook in a preliminary practice run of the simulation for approximately 10 min in order to familiarise themselves with the functionality of the simulation, this familiarisation step was not required for the professional ATCOs considering the ample experience they possessed. This practice run was not as cognitively taxing as the actual vigilance task, but still required careful attention and allowed the student ATCOs to get familiar with the commands needed to relay to the pseudo-pilot. The simulation was administered in the ATMRI's state-of-the-art aerodrome simulator that offers the ability to provide the most naturalistic environment for aerodrome control duties and operations given its ability to simulate the working environments of the Changi Airport Control Tower Cabin including a 360-degree view of the aerodrome (Figs. 2 and 3). The simulator runs on specialised ATM simulation software such as the NLR Air Traffic Control Research Simulator (NARSIM) and AirTOP, two of the simulation tools used for airfield performance assessment and for airport resource utilisation optimisation.

fNIRS System and Procedure
The entirety of the experiment was conducted utilising a continuous-wave fNIRS device first described by Chance et al. (1998), further developed at Drexel University (Philadelphia, PA), manufactured and supplied by fNIR Devices LLC (Potomac, MD). The purpose of the fNIRS device is to continuously monitor the prefrontal cortex or the executive decision-making brain regions, whilst participants complete the vigilance task.

Fig. 2. ATMRI Tower Simulator

Fig. 3. Participant conducing the ATCO Vigilance task in the ATMRI's Tower Simulator. Heavy rain and reduced visibility are introduced into the simulation to provide a more naturalistic setting for multiple missed approaches.

The device is consists of three main components: a headpiece (sensor pad) which is responsible for holding the light sources and detectors, establishing placement for the

16 opcodes: a control box to manage the hardware: and a computer which receives and stores the continuous fNIRS data for further data analysis (Fig. 4).

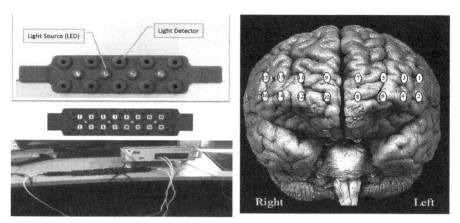

Fig. 4. fNIR sensor with 4 light sources and 10 detectors (left, top) and 16 optode (channel) measurement locations registered on sensor (left, middle) and juxtaposed on brain surface image

The fNIRS sensor provides 10 Hz sampling rate, and a 2.5 cm source-detector separation offering approximately 1.25 cm–1.5 cm penetration depth. The light emitting diodes (LEDs) were activated one light source at a time with the four surrounding photodetectors being sampled. The headpiece containing the sensor is positioned over the forehead providing a total of 16 active optodes or channels, that monitor the dorsal and inferior cortical areas located underneath the forehead (Ayaz et al. 2012; Baunce et al. 2006; Izzetoglu 2005). The difference in light absorption between oxygenated haemoglobin and deoxygenated haemoglobin (HbO_2 and HbR) and is measured by fNIRS by their respective wavelengths as dependent measures versus time by using a modified Beer-Lambert law (Jobsis 1977; Cope and Delpy 1988).

(Bottom right) fNIR Device model 2000s (bottom, left). Brain surface image is from University of Washington, Digital Anatomist Project.

3   Results

Preliminary results from an independent samples t-test indicated that when comparing the oxygenation values underneath channel 14 between the professional ATCO and the novice ATCOs for the first 3 missed approaches, there was a significant difference in oxygenation values for professional ATCOs (M = 7.58, SD = 2.14) vs novice ATCOs (M = 0.17, SD = 3.61); t(16) = 5.10, p = .0002 (Fig. 5). These results suggest when comparing the brain activation between pro ATCOs vs novice ATCOs, there exists a significant difference in activation from the right dorsolateral prefrontal cortex, which has been associated with sustained attention or more precisely, vigilance (Sanchez et al. 2016).

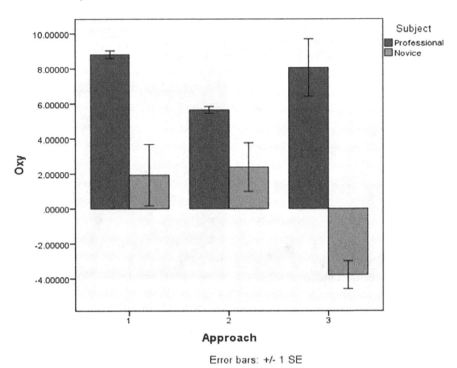

Fig. 5. Comparison of average OXY values between professional and expert ATCOs.

4   Conclusion and Discussion

This study has provided insightful understanding of how fNIRS can be incorporated into realistic, high-fidelity environments to quantitatively measure the performance of operators in their workplaces. Although significant differences existed between the professional vs novice ATCOs in terms of their vigilance activation as reflected in channel 14 representing the right dlPFC, the left dlPFC, which is more associated with spatial working memory was not activated at the same levels. We hypothesise that this could be attributed to the fact that both pro and novice ATCOs seldomly had to take into account where the aircraft was located on the simulation, since they had access to a radar display monitor that showed the aircraft's positioning as it entered their purview. Future studies could specifically inform ATCOs to attenuate to both the simulation screen and their radar display to reflect this spatial working memory task in the left dlPFC. Moreover, fNIRS studies typically include a behavioural component in which the quantitative fNIRS data can be compared to. However, in our study, this was not included due to increased complexity to the experimental protocol. To account for this, All participants had to adhere to the ICAO standard of 5 nautical mile separation; equivalent to approximately one minute and a half of separation between arrivals, thus establishes an uniform behavioural baseline across all participants. Though only 4 ATCO were used for our study, half were professionally trained ATCOs, which negated the need to acquire more participants. Future studies would utilize more ATCOs in order to see if any other differences could emerge from a larger sample size, as well as seeing if our results could be replicated. Furthermore, there were times in which participants would become bored due to the waiting period between arriving flights and would move their heads around, increasing the amount of motion artefacts and reducing the amount of usable data. To circumvent this in the future, we would implement a more robust experimental protocol that would have participants not moving around too much, in order to have more usable data to analyse.

Acknowledgment. This work was made possible through the NTU-CAAS Research Grant M4062429.056.706022 by Air Traffic Management Research Institute, and the school of mechanical engineering NTU. Ethics approval for this research was granted by the Nanyang Technological University (NTU) institutional review board (IRB) (IRB-2020-01-026). A special thank you goes out to Lab Manager Kevin of the ATMRI whose assistance with the tower simulator made the whole project possible.

References

Ayaz, H., et al.: Optical brain monitoring for operator training and mental workload assessment. Neuroimage **59**(1), 36–47 (2012)

Baunce, S., Izzetoglu, M., Izzetoglu, K., Onaral, B., Pourrezae, K.: Functional near-infrared spectroscopy. An emerging neuroimaging modality. IEEE Eng. Med. Biol. Mag. **25**, 54–62 (2006)

Bekier, M., Molesworth, B.R., Williamson, A.: Tipping point: the narrow path between automation acceptance and rejection in air traffic management. Saf. Sci. **50**(2), 259–265 (2012)

Berman, K.F., Weinberger, D.R.: Lateralisation of cortical function during cognitive tasks: regional cerebral blood flow studies of normal individuals and patients with schizophrenia. J. Neurol. Neurosurg. Psychiatry **53**(2), 150–160 (1990)

Billings, C.E.: Aviation Automation: The Search for a Human-Centered Approach. CRC Press, Boca Raton (2018)

Boeing Commercial Airplanes: Statistical summary of commercial jet airplane accidents (2018)

Clark, V.P., et al.: TDCS guided using fMRI significantly accelerates learning to identify concealed objects. Neuroimage **59**(1), 117–128 (2012)

Coffman, B.A., Clark, V.P., Parasuraman, R.: Battery powered thought: enhancement of attention, learning, and memory in healthy adults using transcranial direct current stimulation. Neuroimage **85**, 895–908 (2014)

Coffman, B.A., et al.: Impact of tDCS on performance and learning of target detection: interaction with stimulus characteristics and experimental design. Neuropsychologia **50**(7), 1594–1602 (2012)

Cohen, R.M., Semple, W.E., Gross, M., King, A.C., Nordahl, T.E.: Metabolic brain pattern of sustained auditory discrimination. Exp. Brain Res. **92**(1), 165–172 (1992)

Cope, M., Delpy, D.T.: System for long-term measurement of cerebral blood and tissue oxygenation on newborn infants by near infra-red transillumination. Med. Biol. Eng. Compu. **26**(3), 289–294 (1988)

Coull, J.T., Frackowiak, R.S., Frith, C.D.: Monitoring for target objects: activation of right frontal and parietal cortices with increasing time on task. Neuropsychologia **36**(12), 1325–1334 (1998)

Curtin, A., Ayaz, H.: The age of neuroergonomics: towards ubiquitous and continuous measurement of brain function with fNIRS. Jpn. Psychol. Res. **60**(4), 374–386 (2018)

Dai, L., Liu, Y., Hansen, M.: Modelling Go-Around Occurrence (2019)

Derosière, G., Mandrick, K., Dray, G., Ward, T.E., Perrey, S.: NIRS-measured prefrontal cortex activity in neuroergonomics: strengths and weaknesses. Front. Hum. Neurosci. **7**, 583 (2013)

Dockery, C.A., Hueckel-Weng, R., Birbaumer, N., Plewnia, C.: Enhancement of planning ability by transcranial direct current stimulation. J. Neurosci. **29**(22), 7271–7277 (2009)

Falcone, B., Coffman, B.A., Clark, V.P., Parasuraman, R.: Transcranial direct current stimulation augments perceptual sensitivity and 24-hour retention in a complex threat detection task. PLoS ONE **7**(4), e34993 (2012)

Federal Aviation Administration Annual Performance Report. Washington DC, USA (2015)

Federal Aviation Administration: Federal Aviation Regulations/Aeronautical Information Manual 2014. Skyhorse Publishing, New York (2013)

Fedota, J., Parasuraman, R.: Neuroergonomics and human error. Theor. Issues Ergon. Sci. **11**, 402–421 (2010)

Ferrari, M., Mottola, L., Quaresima, V.: Principles, techniques, and limitations of near infrared spectroscopy. Can. J. Appl. Physiol. **29**(4), 463–487 (2004)

Fregni, F., et al.: Anodal transcranial direct current stimulation of prefrontal cortex enhances working memory. Exp. Brain Res. **166**(1), 23–30 (2005)

Izzetoglu, K., Richards, D.: Human performance assessment: evaluation of wearable sensors for monitoring brain activity. In: Vidulich, M., Tsang, P. (eds.) Improving Aviation Performance through Applying Engineering Psychology: Advances in Aviation Psychology, 1st edn, pp. 163–180. CRC Press, Boca Raton (2019)

Izzetoglu, K., Bunce, S., Onaral, B., Pourrezaei, K., Chance, B.: Functional optical brain imaging using near-infrared during cognitive tasks. Int. J. Hum.-Comput. Interact. **17**(2), 211–227 (2004)

Izzetoglu, M.: Functional near-infrared neuroimaging. IEEE Trans. Neural Syst. Rehabil. Eng. **13**(2), 153–159 (2005)

Jobsis, F.F.: Noninvasive, infrared monitoring of cerebral and myocardial oxygen sufficiency and circulatory parameters. Science **198**(4323), 1264–1267 (1977)

Kirwan, B.: The role of the controller in the accelerating industry of air traffic management. Saf. Sci. **37**(2–3), 151–185 (2001)

Kocsis, L., Herman, P., Eke, A.: The modified Beer-Lambert law revisited. Phys. Med. Biol. **51**(5), N91 (2006)

Leveson, N., et al.: A safety and human-centered approach to developing new air traffic management tools. In: Proceedings Fourth USA/Europe Air Traffic Management R&D Semina, pp. 1–14. December 2001

Lewin, J.S., et al.: Cortical localization of human sustained attention: detection with functional MR using a visual vigilance paradigm. J. Comput. Assist. Tomogr. **20**(5), 695–701 (1996)

Liu, Y., Wickens, C.D.: Mental workload and cognitive task automaticity: an evaluation of subjective and time estimation metrics. Ergonomics **37**(11), 1843–1854 (1994)

McKendrick, R., Parasuraman, R., Ayaz, H.: Wearable functional near infrared spectroscopy (fNIRS) and transcranial direct current stimulation (tDCS): expanding vistas for neurocognitive augmentation. Front. Syst. Neurosci. **9**, 27 (2015)

Milton, J., Solodkin, A., Hluštík, P., Small, S.L.: The mind of expert motor performance is cool and focused. Neuroimage **35**(2), 804–813 (2007)

Nelson, J.T., McKinley, R.A., Golob, E.J., Warm, J.S., Parasuraman, R.: Enhancing vigilance in operators with prefrontal cortex transcranial direct current stimulation (tDCS). Neuroimage **85**, 909–917 (2014)

Obrig, H., Villringer, A.: Beyond the visible- imaging the human brain with light. J. Cereb. Blood Flow Metab. **23**(1), 1–18 (2003)

Parasuraman, R.: Neuroergonomics: Brain, cognition, and performance at work. Curr. Dir. Psychol. Sci. **20**(3), 181–186 (2011)

Parasuraman, R. (ed.): The Attentive Brain. Mit Press, Cambridge (2000)

Parasuraman, R., McKinley, R.A.: Using noninvasive brain stimulation to accelerate learning and enhance human performance. Hum. Factors **56**(5), 816–824 (2014)

Pardo, J.V., Fox, P.T., Raichle, M.E.: Localization of a human system for sustained attention by positron emission tomography. Nature **349**(6304), 61 (1991)

Redding, R.E.: Analysis of operational errors and workload in air traffic control. In: Proceedings of the Human Factors and Ergonomics Society Annual Meeting, vol. 36, no. 17, pp. 1321–1325. SAGE Publications, Los Angeles (1992)

Rodgers, M.: Human Factors Impacts in Air Traffic Management. Routledge, London (2017)

Rueckert, L., Grafman, J.: Sustained attention deficits in patients with right frontal lesions. Neuropsychologia **34**(10), 953–963 (1996)

Sanchez, A., Vanderhasselt, M.A., Baeken, C., De Raedt, R.: Effects of tDCS over the right DLPFC on attentional disengagement from positive and negative faces: an eye-tracking study. Cogn. Affect. Behav. Neurosci. **16**(6), 1027–1038 (2016)

Seamster, T.L., Redding, R.E., Cannon, J.R., Ryder, J.M., Purcell, J.A.: Cognitive task analysis of expertise in air traffic control. Int. J. Aviation Psychol. **3**(4), 257–283 (1993)

Westin, C., Borst, C., Hilburn, B.: Strategic conformance: overcoming acceptance issues of decision aiding automation? IEEE Trans. Huma. Mach. Syst. **46**(1), 41–52 (2015)

A Multi-stage Theory of Neurofeedback Learning

Eddy J. Davelaar$^{(\boxtimes)}$ (ID)

Birkbeck, University of London, London WC1E 7HX, UK
e.davelaar@bbk.ac.uk

Abstract. Neurofeedback is a training paradigm through which trainees learn to voluntarily influence their brain dynamics. Recent years have seen an exponential increase in research interest into this ability. How neurofeedback learning works is still unclear, but progress is being made by applying models from computational neuroscience to the neurofeedback paradigm. In this chapter, I will present a multi-stage theory of neurofeedback learning, which involves three stages, involving different neural networks. In stage 1, the system discovers the appropriate goal representation for increasing the frequency of positive feedback. This stage operates at a within-session timescale and is driven by reward-based learning, which updates fronto-striatal connections. Stage 2 operates on a timescale that covers multiple training sessions and is sensitive to consolidation processes. This stage involves updating striatal-thalamic and thalamo-cortical connections. Finally, after stages 1 and 2 have started, stage 3 may be triggered by the awareness of the statistical covariation between interoceptive and external feedback signals. When this awareness emerges, neurofeedback learning may speed up and its effect be maintained well after the conclusion of the training period. Research guided by this framework is described that consist of quantitative, qualitative, and computational methodologies. At present, the findings suggest that the framework is able to provide novel insights into the nature of neurofeedback learning and provides a roadmap for developing instructions that are designed to facilitate the likelihood of learning success.

Keywords: Neurofeedback · Computational neuroscience · EEG · Neuroimaging

1 Introduction

In the early 1900s, psychological research was dominated by a methodological approach championed by behaviorism that looked into the smallest components that could explain behavior. This approach led to discoveries of classical and operant conditioning that could explain behaviors observed in animals and humans. It also influenced thinking about education, parenting, and advertising. A lesser known fact is that conditioning principles were not only investigated at the level of overt behaviors, such as pressing a lever or pecking at a light, but work by a range of researchers uncovered

© Springer Nature Switzerland AG 2020
D. D. Schmorrow and C. M. Fidopiastis (Eds.): HCII 2020, LNAI 12196, pp. 118–128, 2020.
https://doi.org/10.1007/978-3-030-50353-6_9

that brain oscillations could be conditioned as well [1–3]. This work was the foundation of a field called neurofeedback in which the brain's activation is modified through conditioning. Neurofeedback was mainly focused on electroencephalography (EEG) for many decades, but over the last 20 years and especially over the recent 5 years a range of other brain measurement techniques have been used, such as magnetoencephalography (MEG, [4, 5]), functional magnetic resonance imaging (fMRI, [6, 7]), and functional near-infrared spectroscopy (fNIRS, [8, 9]). This increased interest brings with it a need to understand the biological mechanisms underlying neurofeedback learning. In addition, whilst neurofeedback alters the brain activation, it has been argued that it also leads to functional changes in performance and subjective experience. This is the foremost reason that neurofeedback has a long history as a neurotherapeutic intervention for psychological conditions, with epilepsy and attention deficit hyperactivity disorder, as two of the earlier conditions for which therapeutic benefits were recorded. Understanding the mechanisms underlying neurofeedback learning will benefit clinicians in improving their success rate and enhance hypothesis-driven research. This chapter will review some of the early literature on conditioning of EEG oscillations. This is followed by a discussion on the current state of research, highlighting the various methodological challenges. A multi-stage theory of neurofeedback learning is then introduced, with each stage addressed in greater detail. Finally, the theoretical and practical implications of this model will be explicated.

2 Conditioning of EEG Oscillations: The Early Work

Behaviorist research in the early 1900s was focused on tabulating the smallest association that could lead to an overt behavior. As mentalist topics such as attention and memory were not directly observable, these were not regarded as acceptable areas of inquiry. What was acceptable was measuring any physiological variable and check whether it could be conditioned. In a largely forgotten literature, a particular focus was whether stimuli presented to any sense organ could become a conditioned stimulus for what is called alpha blocking.

2.1 Classical Conditioning

Alpha blocking is the phenomenon that the power of the alpha oscillation over the visual areas decreases when the participant opens the eyes. This phenomenon was considered a natural reflex. Several studies addressed whether this reflex could be conditioned to stimuli other than light. Jasper and Shagass [1] conducted an extensive investigation with sound. Their participants were on a bed in a darkened room with an electrode attached over the right occipital area. They were tasked with pressing a response button as soon as they saw a light. Pressing the button did not have any effect of the alpha oscillation. In conditioning trials, the light was preceded by a tone. This tone became the conditioned stimulus, as demonstrated by the decrease of alpha power after playing the tone, but without illuminating the light. The simple conditioning occurred quite quickly, but was also rapidly extinguished.

Jasper and Shagass [1] investigated the conditioned alpha block using simple, cyclic, delayed, trace, differential, differential delayed, and backward conditioning, thereby establishing that "higher centres, not necessarily involving peripheral effector systems" (p. 384) can be Pavlovian conditioned. An interesting side note is that during the extinction period, spontaneous recovery of the conditioned response can occur, which indicates that not only that the tone got associated with the presence of light, but also with its absence during extinction. The spontaneous recovery occurs when the relative strength of the former outweighs the latter.

In a follow-up study, Jasper and Shagass [2] asked participants to subvocally say "block" and press a button and keep it pressed until subvocally saying "stop" and releasing the button. The timing of the subvocalisation was entirely voluntary. During conditioning trials, the experimenter allowed the electrical circuit to switch on a light when the participant depressed the button. On test and extinction trials, the light remained switched off. A conditioned response was observed in this scenario in the absence of an external stimulus. This study demonstrated that a conscious mental act was able to become a conditioned stimulus.

In both involuntary and voluntary studies of the alpha block, there was rapid extinction of the conditioned response. Nevertheless, it highlighted that the brain and its higher centres follow the same rules of conditioning as overt behaviors, such as the salivation reflex.

2.2 Operant Conditioning

Whereas the conditioned alpha block is considered a Skinner Type II conditioned response, Wyrwicka and Sterman [3] demonstrated that brain oscillations can be conditioned through operant conditioning. In their study, they had cats who were deprived of food for 22 h. They received condensed milk whenever they exhibited a burst of sensorimotor rhythm (SMR) over the sensorimotor cortex for at least 0.5 s. The cats were able to increase the occurrence of SMR. Of particular interest are the visually recorded behaviors that the cats engaged in. All cats converged on a different posture that can be described as freezing or staring. Immediately after the milk was consumed the cats returned and adopted the same posture. In addition, when after an extinction period a reconditioning phase started, the cats returned again to their individual posture. Thus, not only was the SMR oscillation subject to operant conditioning, it also coincided with a specific behavior that took different forms. In lay terms, it is as if the cats had to "go into their zone", after which SMR developed in a few seconds.

Similar correlations have been observed in other studies with human participants. For example, in alpha training participants report different phenomenology [10, 11], which suggests that changes in the brain activation profile influences the subjective experience. However, it is yet unclear whether the subjective experience is shared among individuals or idiosyncratic. Some initial work in this direction is currently being conducted [12, 13].

3 Neurofeedback Research: Current Developments

Despite over eight decades of research in neurofeedback, the field is currently at a crossroads. Most of the research is conducted using EEG and has spawned several debates, such as whether the conditioned alpha block is actually reflecting sensitization and whether clinical trials are appropriately placebo-controlled. The demonstration of successful neurofeedback of the BOLD signal has opened up a much wider field with its own technical challenges. Together with a lack of theoretical framework for generating hypotheses, the consequence has been that EEG neurofeedback is still being considered as flawed.

At the time of writing, there are major developments afoot that rehabilitate EEG neurofeedback. Dedicated special issues on the topic feature many methodological and technical advances that were not available two decades ago. In addition, dedicated software for research purposes are being developed, some are expected to be Open Accessible. Sharing of data through the Open Science Framework and pre-registration of studies are being considered and implemented.

Although the challenges of the research environments are being met, the theoretical developments are still in need of major work. General high-level descriptions have been proposed to provide a bird's eye view of neurofeedback. However, generating testable hypotheses from these perspectives has remained elusive. The proposal in this chapter is that models from computational neuroscience could be utilized to develop a mechanistic understanding of neurofeedback learning. These models could then be used to implement new research designs and generate hypotheses. They can be used to test some of the higher-level descriptions of neurofeedback learning, thereby allowing comparison of different theoretical viewpoints.

4 A Multi-stage Theory of Neurofeedback Learning

The empirical research base is rich and vast enough for developing formal theories to further drive the field forward. Unfortunately, the marriage between theorists within computational neuroscience and researchers in applied neuroscience never took hold. When taking a computational neuroscience approach to neurofeedback, insights can be gained that were not obvious at first. The multi-stage theory of neurofeedback learning [14] is a product of merging the two disciplines.

4.1 Overview of the Theory

The theory assumes three stages that involve different neural networks (see Fig. 1). In stage 1, the system discovers the appropriate goal representation for increasing the frequency of positive feedback. This stage operates at a within-session timescale and is driven by reward-based learning, which updates fronto-striatal connections. Stage 2 operates on a timescale that covers multiple training sessions and is sensitive to consolidation processes that unfold during sleep. This stage involves updating striatal-thalamic and thalamo-cortical connections. In effect, this stage changes the set point of the system, making it easier to produce the target brain oscillation. Finally, after stages 1 and 2 have started, stage 3 may be triggered by the awareness of the statistical covariation between interoceptive and external feedback signals. When this awareness emerges, neurofeedback learning may speed up and its effect be maintained well after the conclusion of the training period.

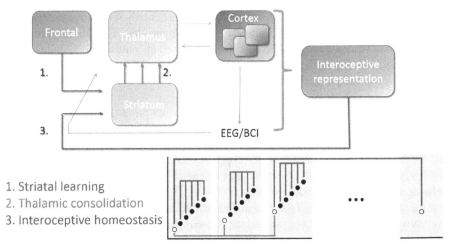

Fig. 1. The multi-stage theory of neurofeedback learning as applied to EEG neurofeedback. The BCI system records the EEG oscillations and convert this into a feedback signal. This signal is used in reinforcement learning through which the frontal goal representation gets associated with neural patterns over the striatum that lead to more positive feedback. This positive feedback loop is assumed to underlie within-session learning curves. The second stage starts after the first and involves updating the striatal-thalamic connections. As this unfolds over a longer time-scale, it is assumed to underlie the learning curves over sessions. Brain patterns may correspond with unique subjective experiences. When these exist for the target brain pattern, they make become secondary reinforcers and this stage is assumed to underlie self-reinforcement in the absence of external feedback and maintenance of the acquired skill.

As the stages operate at different time-scales, learning curves within and between sessions are hypothesized to reflect these different stages. In addition, the implication is that in research a positive learning curve could be observed within, but not between sessions. This would not mean that no neurofeedback learning occurred. Instead, it could mean that stage 2 does not occur for that neurofeedback protocol, as would be the case if changing the setpoint is a physiological impossibility. A closer look at learning curves is warranted to scrutinize their relation to particular protocols.

At present, the theory is mainly a framework in which to explain neurofeedback findings. However, as additional data is being addressed, gaps in the theory will inevitably become visible, which require dedicated hypothesis-driven research. This is the advantage of a detailed theory. To facilitate the identification of directions for further inquiry, each stage will now be addressed in more detail with presentation of work that was inspired by the theory.

4.2 Stage 1

Stage 1 assumes that during learning, a frontal representation that contains the person's goal (i.e., increase the number of reward signals) is associated with a random neural pattern over the striatum that increases the likelihood of reward. In the original article introducing the theory, the frontal and striatal parts were analyzed in computational neuroscience and in a mathematical model. The task for this sub-model was to move from a state of producing the baseline EEG pattern to a state in which the target brain pattern (in that case alpha oscillations) was more likely. By implementing basic equations of reinforcement learning, the models were shown to be able to learn, demonstrating that computationally stage 1 of the theory is indeed possible.

The model produced two new insights. First, the space of possible striatal patterns is immense and finding the target pattern through trial-and-error is highly unlikely. The updates to the fronto-striatal connections interact with the strong intra-striatal inhibition to implement a selection mechanism that drives the system to converge on a stable pattern. Thus, neurofeedback learning in stage 1 is a search process. The convergence counters the positive feedback loop, making stage 1 a self-limiting process. In other words, learning in stage 1 will eventually stop. This insight has repercussions on the evaluation of learning success. In particular, learning success could be defined as a linear increase, as is typically done in the literature, or when an asymptotic level is nearly reached. These are different parts of a sigmoidal learning curve.

The second insight, based on mathematical analysis, is that the probabilistic nature of the EEG generation implies that not all target states are rewarded. This fundamentally changes the neurofeedback paradigm from the often assumed continuous reinforcement schedule (i.e., every target state is rewarded) to a variable intermittent schedule. This is due to the neurofeedback software to reward based on the EEG pattern, whereas the target state produces both the target pattern and the nontarget pattern. This has a consequence that target state is both rewarded and not rewarded in the same training session, which slows down overall learning.

The learning in the stage 1 is particularly sensitive to the threshold setting that is used to decide whether to provide rewards. Set the threshold too low and the system does not learn. Set too high the system unlearns. Yet, the model was used to demonstrate that changing the threshold as a function of the preceding recordings lead to steeper learning curves with higher asymptotic levels [15].

The stage-1 submodel challenges and extends a range of methodological choices that could be further explored. In addition, the interpretation of stage-1 learning as a search process influenced by threshold settings allows a careful consideration of how to devise algorithms for optimal thresholding. Finally, it also allows exploring variations in the feedback protocol. That is whether feedback should be binary (e.g., a beep) or continuous (e.g., volume change), be only positive (e.g., addition of points) or also include negative feedback (e.g., subtraction of points). These issues are currently being investigated.

4.3 Stage 2

Stage 2 puts the thalamus at its centre. However, this is not to say that all neurofeedback needs to involve the thalamus. For example, in fMRI neurofeedback of the amygdala, the striatum influences the amygdala response. In EEG neurofeedback, the thalamus is at the centre of brain oscillations. Addressing the hypothesis of thalamic consolidation or changing the setpoint was approached in a pure computational manner by quantitatively fitting a biophysical model of thalamocortical interactions to EEG data obtained before, during, and after neurofeedback training.

The model that was utilized was developed by Robinson and colleagues [16]. This model takes a mean field approach and contains a wide range of parameters that are constrained from neurophysiological data. This particular model has been solved by the authors and shown to quantitatively fit actual EEG spectra by changing neurophysiological parameters, such as intrathalamic, thalamocortical, and cortico-cortical connectivity. Applying this model to actual data allows evaluating whether thalamic connections are critical in understanding the change in EEG oscillations.

The data comes from a study in which participants were trained to increase alpha and theta over electrode Pz over the course of ten training sessions. Participants had their eyes closed and focused on the sounds of a babbling brook (alpha) and of the ocean (theta) for fifteen minutes. Their task was to increase the volume of both sounds. This particular protocol is known for a phenomenon called the alpha/theta crossover, whereby after a period of higher alpha power compared to theta power, a switch occurs where theta dominates the power spectrum. As part of an ongoing investigation, individual sessions were checked for the crossover pattern. This was found for one person in the tenth session (see Fig. 2).

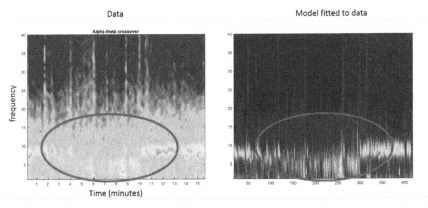

Fig. 2. Time-frequency spectra demonstrating the alpha/theta crossover (inside the red ovals). Left panel: Actual data. Right panel: Model fitted to the data. (Color figure online)

The biophysical model was fitted to the data and the values of the parameters were plotted against time (see Fig. 3). This fitting routine requires updating several parameters of which five are shown in Fig. 3. The dendritic rate constant relates to the rate of processing in the dendritic tree of pyramidal cortical cells. The inhibitory and excitatory gains relate to the cortico-cortical connections. These three parameters relate to cortical neurons only. The two thalamic parameters are the negative thalamic gain, which is composed of the pathway coming from the cortex to the reticular nucleus to the thalamic relay neurons and then back to the cortex. The negative feedback loop between the relay neurons and the reticular nucleus is labelled here as the intra-thalamic gain. The positive thalamic loop is not shown here.

The alpha/theta crossover period is clearly reflected in the decreased cortical excitatory gain. However, none of the three cortical parameters predict the ensuing cross-over. Both thalamic parameters show a gradual decrease in negative gain before the crossover

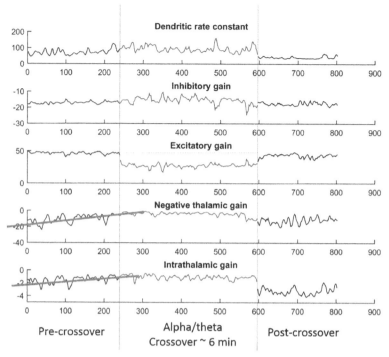

Fig. 3. Parameter values as a function of time in the pre-crossover, crossover and post-crossover periods. The two thalamic parameters seem to become less negative before the crossover period. A sudden switch seems to occur whereby cortical excitatory gains drop to a level that is balanced by the inhibitory cortical gain.

period, remain constant during this period and drop back to pre-crossover levels after the period. Although more work is certainly needed in this area, the model fits suggest that the decrease in the inhibitory influence of the reticular nucleus could trigger the alpha/theta crossover. In the multi-stage theory this would be consistent with increased inhibition from the basal ganglia to the reticular nucleus during training.

4.4 Stage 3

Stage 3 of the multistage theory assumes that patterns of brain oscillations covary with subjective experiences. This finding supports other work within the field of neurophenomenology [17] and converges with demonstrations of differential experiences in fMRI neurofeedback [12]. In EEG neurofeedback there is evidence that participants scoring high on introspective ability show a greater difference in alpha (at Oz) duration between alpha generation and alpha suppression periods than those who score low of introspective ability [11]. Therefore, it is not only possible that during neurofeedback training, participants become aware of sensations that could be used in further facilitating learning, but that being a meditator or having mind-body awareness allows better control over brain activations [18, 19].

To test this link we [13] analyzed verbal reports of participants that completed a single session of frontal alpha training. As not all participants managed to increase their alpha, we divided the sample into two groups: learners and non-learners. This classification was based on the EEG spectral power over the training period. After grouping the participants the verbal reports were examined for group differences. Figure 4 presents a summary of the results. Learners compared to non-learners were more aware of themselves and the environment, whereas non-learners compared to learners were preoccupied with trying things out. Even trying to relax was not helpful in increasing alpha. These findings have been converted into instructions and given to a new set of participants who were either in a true neurofeedback training session or a sham-control condition. Preliminary results show that the instructions do facilitate neurofeedback learning. This two-phase research design (i.e., mixed-method study followed by an instruction-implementation study) provides a blueprint for developing instruction that are duly tested in training studies.

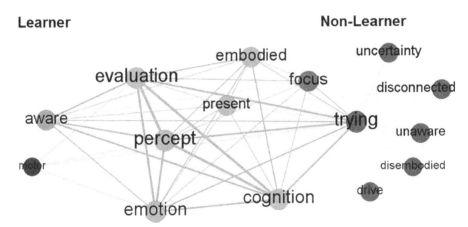

Fig. 4. Network visualization of the topics reported by learners and non-learners. The size of each node is proportional to the number topic-instances. The thickness of the inter-node connections reflect the frequency with which the two topics are present in verbal reports. Red nodes are topics that are reported more often by non-learners compared to learners and vice-versa for the blue nodes. Reproduced from [13]. (Color figure online)

5   Implications for Research and Practice

The multistage theory of neurofeedback learning is a theory that is sufficiently specific to test its assumptions and yet open enough to connect with other theories and methodologies. Most of the implications from this theory is for researchers. Many assumptions in neurofeedback research are hidden, either in the methods of thresholding the interpretation of learning curves, and even whether neurofeedback requires consciousness. The framework allows putting these questions and in doing so improve the framework or develop a better one. This process of theoretical development has been lacking in the neurofeedback literature.

For neurofeedback practitioners, the model provides a number of directions through which to augment clinical practice. First of all, providing as much information to clients and their carers is vital for building mutual trust and facilitating therapy adherence. Secondly, software can be developed that track the thalamocortical connectivities over the course of the training programme. It would augment a clinical assessment report that already includes a quantitative EEG (QEEG) component – a summary of the power in all frequency bands for all electrode placements. Whereas the QEEG provides a description of the brain oscillations, parameter plots provide additional information about the latent neurophysiological parameters.

The theory is still in its infancy and parts are still being developed. As brain-computer interfaces become more common, understanding how they work and how to facilitate learning success is key to develop practical applications that yield replicable results.

References

1. Jasper, H., Shagass, C.: Conditioning the occipital alpha rhythm in man. J. Exp. Psychol. **28**(5), 373–388 (1941)
2. Jasper, H., Shagass, C.: Conscious time judgments related to conditioned time intervals and voluntary control of the alpha rhythm. J. Exp. Psychol. **28**(6), 503–508 (1941)
3. Wyrwicka, W., Sterman, M.B.: Instrumental conditioning of sensorimotor cortex EEG spindles in the waking cat. Physiol. Behav. **3**(5), 703–707 (1968)
4. Florin, E., Bock, E., Baillet, S.: Targeted reinforcement of neural oscillatory activity with real-time neuroimaging feedback. Neuroimage **88**, 54–60 (2014)
5. Okazaki, Y.O., Horschig, J.M., Luther, L., Oostenveld, R., Murakami, I., Jensen, O.: Real-time MEG neurofeedback training of posterior alpha activity modulates subsequent visual detection performance. Neuroimage **107**, 323–332 (2015)
6. Sulzer, J., Haller, S., Scharnowski, F., Weiskopf, N., Birbaumer, N., Blefari, M.L., et al.: Real-time fMRI neurofeedback: progress and challenges. Neuroimage **76**, 386–399 (2013)
7. Emmert, K., Kopel, R., Sulzer, J., Bruhl, A.B., Berman, B.D., Linden, D.E.J., et al.: Meta-analysis of real-time fMRI neurofeedback studies using individual participant data: how is brain regulation mediated? Neuroimage **124**, 806–812 (2016)
8. Sakatani, K., Takemoto, N., Tsujii, T., Yanagisawa, K., Tsunashima, H.: NIRS-based neurofeedback learning systems for controlling activity of the prefrontal cortex. Adv. Exp. Med. Biol. **789**, 449–454 (2013)
9. Marx, A.-M., Ehlis, A.-C., Furdea, A., Holtmann, M., Banaschewski, T., Brandeis, D., et al.: Near-infrared spectroscopy (NIRS) neurofeedback as a treatment for children with attention deficit hyperactivity disorder (ADHD)—a pilot study. Front. Human Neurosci. **8**, 1038 (2015)
10. Nowlis, D.P., Kamiya, J.: The control of electroencephalographic alpha rhythms through auditory feedback and the associated mental activity. Psychophysiology **6**, 476–83 (1970)
11. Ancoli, S., Green, K.F.: Authoritarianism, introspection, and alpha wave biofeedback training. Psychophysiology **14**(1), 40–44 (1977)
12. Garrison, K.A., Santoyo, J.F., Davis, J.H., Thornhill, T.A., Kerr, C.E., Brewer, J.A.: Effortless awareness: using real time neurofeedback to investigate correlates of posterior cingulate cortex activity in meditators' self-report. Front. Hum. Neurosci. **7**, 440 (2013)
13. Davelaar, E.J., Barnby, J.M., Almasi, S., Eatough, V.: Differential subjective experiences in learners and non-learners in frontal alpha neurofeedback: piloting a mixed-method approach. Front. Hum. Neurosci. **12**, 402 (2018)

14. Davelaar, E.J.: Mechanisms of neurofeedback: a computation-theoretic approach. Neuroscience **378**, 175–188 (2018)
15. Davelaar, E.J.: A computational approach to developing cost-efficient adaptive-threshold algorithms for EEG neurofeedback. Int. J. Struct. Comput. Biol. **1**, 1–4 (2017)
16. Rowe, D., Robinson, P., Rennie, C.: Estimation of neurophysiological parameters from waking EEG using a biophysical model of brain dynamics. J. Theor. Biol. **231**(3), 413–433 (2004)
17. Varela, F.: Neurophenomenology: a methodological remedy to the hard problem. J. Conscious. Stud. **3**, 330–350 (1996)
18. Ta, L.F., Dienes, Z., Jansari, A., Goh, S.Y.: Effect of mindfulness meditation on brain-computer interface performance. Conciousness Cogn. **23**, 12–21 (2014)
19. Cassady, K., You, A., Doud, A., He, B.: The impact of mind-body awareness training on the early learning of a brain-computer interface. Technology **2**(3), 254–260 (2014)

Analyses of Impression Changes and Frontal Lobe Activity While Viewing Videos

Anna Endo[1], Naoki Takahashi[1], Takashi Sakamoto[2(✉)], and Toshikazu Kato[1]

[1] Chuo University, 1-13-27 Kasuga, Bunkyo, Tokyo 112-8551, Japan
a15.fark@g.chuo-u.ac.jp, {naoki,t-kato}@kc.chuo-u.ac.jp
[2] National Institute of Advanced Industrial Science and Technology (AIST),
Tsukuba, Ibaraki 305-8568, Japan
takashi-sakamoto@aist.go.jp

Abstract. This study investigates the mechanism of emotional processing in the human brain. We analyzed the association between impressions perceived from videos and brain activity in the prefrontal cortex. In particular, we focused on the differences between impressions of videos in the first-time viewing and second time viewing. The participants' brain activities were measured using optical topography equipment that provided near-infrared spectroscopy (NIRS). The experimental results revealed that changes in the perceived impressions were strongly correlated with brain activity in the dorsolateral prefrontal cortex (DLPFC). We discovered that the brain activities in the DLPFC were lower when the impressions of the videos in the first-time viewing were evaluated as "good" by the participants. We also established that brain activities outside the prefrontal cortex were more active in the second time viewing. Our research results on video impressions and brain activities will contribute, in the near future, to the development of, among other areas, brain-machine interfaces, neuromarketing, and affective computing.

Keywords: NIRS · Impression · Affective engineering · Brain activities · Prefrontal cortex

1 Introduction

The purpose of this study was to analyze the association between image impressions and the prefrontal cortex activity to clarify the generation mechanism of kansei, such as impressions, feelings, emotions, intuition, or preference. It is especially important to clarify the mechanism of human qualitative factors such as impressions. Recently, the consideration of impressions and emotions has become an important research matter in human-computer interaction and affective computing [1]. It is expected that a basic knowledge of emotional processing in the brain will be applied to human-computer interaction and affective computing.

For example, the research and development of methods for quantitatively measuring individual impressions and emotions enables the quantitative measurement of a consumer's unconscious psychological processes. These techniques are called "neuro-marketing" techniques, which are often used to grasp the qualitative needs of consumers.

© Springer Nature Switzerland AG 2020
D. D. Schmorrow and C. M. Fidopiastis (Eds.): HCII 2020, LNAI 12196, pp. 129–141, 2020.
https://doi.org/10.1007/978-3-030-50353-6_10

Moreover, this technique enables us to read the unconscious emotions of consumers and the process of its change from biological information [2]. Usually, questionnaire surveys and interviews are used to determine consumer needs. These capture only the outcome of decisions; however, we usually make decisions through various psychological processes. Therefore, it is difficult to discern the psychological processes quantitatively in these cases [3]. This problem can be addressed via the quantitative measurement of individual impressions and emotions, such as in neuromarketing.

The biggest concern related to this research is that the association between impressions and the observer's brain is still unclear. To explain this, Azehara and Kato [4] examined the association between the first impression of images and frontal lobe activity. Tanida and Kato [5] investigated the association between sex differences and frontal lobe activity when viewing first-person videos. Following these studies, our research group aimed to investigate the association between image impressions and frontal lobe activity, and analyzed the changes in viewing frequency, impressions, and frontal lobe activity.

Hughes et al. [6] reported on 'good' and 'bad' episodes after presenting human faces and affiliations; observers often downplayed negative information they learned about in-group members. There was significant activity in the dorsolateral prefrontal cortex (DLPFC) in the processing of negative information.

McClure et al. [7] measured the difference in brain activity between participants who were presented with the Coca-Cola brand and those who were presented with the Pepsi brand. As a result, the hippocampus and the DLPFC responded remarkably only when Coca-Cola was shown, and consumed. The participants unconsciously said, "Coca-Cola is delicious." The authors concluded that the hippocampus, which is responsible for memory, and the DLPFC, which is responsible for judgment and processing, work together to determine whether a person likes or dislikes a certain taste by comparing the brand value with the taste, saying, "From past memories (Experience), the participants liked the brand, so the taste of Coca-Cola is better."

As for neuromarketing, Johansson et al. [8] presented two photographs of a person side by side, and let the examinee choose the attractive one. The participants were then given the selected photographs and asked to provide the reason for their choice. Subsequently, they gave them the photographs (of the same person) which were not selected and they were asked to provide the reason for that choice. As a result, 74% of the participants did not notice it. The factors at the time of decision making and their subsequent interpretation do not always correspond.

These studies predict that the hippocampus and the DLPFC are involved in impression determination. However, the relationship between memory and impression has not been considered in depth. With regards to the study by McClure et al. [7], it is necessary to clarify whether impressions are changed by memories and to observe the brain activity. Consequently, this study focuses on a method of near-infrared spectroscopy (NIRS) to evaluate brain activity in the DLPFC. We investigated how impressions and activity of the prefrontal area change when the same image is viewed twice.

2 Methods

2.1 Selection of Videos

First, images used for cerebral function evaluation were selected. The brain function was later evaluated using the image selected. Although omitted in this summary, participants were asked to view 22 videos and answer a questionnaire to calculate a Z-score. Except for monochrome images, the Z-scores were in descending order as follows: Excellent 1, 2, and 3; Normal 1; and Poor 1, 2, and 3.

2.2 Participants

The participants of selection of videos comprised ten men and seven women (mean = 19.2, SD = 1.1). The participants of the measurement of brain activity experiment comprised nine right-handed men and six women (mean = 22.3, SD = 0.4) for cerebral function evaluation.

2.3 Measurement of Brain Activity

In the cerebral function evaluation experiment, the cerebral blood flow was measured using optical topography (ETG-4000 Hitachi Medical) that provided NIRS in order to quantify the psychological activity of the participant. From this change, the active site could be identified.

Optical topography records changes in the levels of oxygenated hemoglobin (oxy-Hb) and deoxygenated hemoglobin (deoxy-Hb) in the cerebral blood stream as time series data. Near-infrared light irradiated to the surface of the head reaches the cerebral cortex as it is absorbed and diffused. The light is then converged again by an optical fiber on the surface of the head. Using the fact that the absorption spectra of oxy-Hb and deoxy-Hb are different, the degree of change in each hemoglobin type is determined [9]. An increase in oxy-Hb levels indicates brain activity. Therefore, we focused on changes in oxy-Hb in this study.

Positron emission tomography (PET) and magnetic resonance imaging (MRI) were also used to measure brain activity. These devices are used in experiments with the subject's body restrained, making it difficult to measure brain activity in a natural state. By contrast, optical topography has the advantage of being able to measure brain activity noninvasively without restricting the position or orientation of the body.

Optical topography was applied to the participant's prefrontal cortex. Numbers 1 to 52 in Fig. 1 indicate the serial numbers of the 52 channels (Ch.) to be measured. The light-receiving probes between Ch. 47 and 48 were located at the frontal pole zero (Fpz), defined by the International 10–20 Act, which is the standard for electroencephalogram electrode arrangement (Fig. 1).

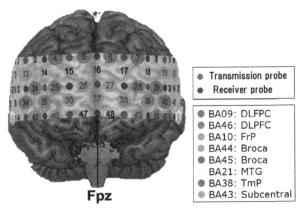

Fig. 1. List of optical topography locations and correspondence between the channels and the prefrontal area

2.4 Stimulation

Participants viewed nine videos selected as one block in the following five steps:

1. image presentation, with a fixation point for 15 s, for stabilization of brain activity;
2. control task video presentation for 30 s;
3. the same image presentation as in step 1, repeated;
4. another video presentation, selected from nine videos, for 30 s; and
5. impression evaluation of the video by hand sign.

These results were the first viewing results. Subsequently, participants viewed the nine videos again. The later results were the second viewing results. The order of the first and second views of the video was randomized.

The videos used were Japanese television commercials broadcasted in Japan [10] that the participants had not seen before. The control task (CT), which was equal to the physical quantity of a video used in the fourth step, was designed to examine the brain activity induced by the image. The CT was a noise video in which pixel values were randomized for each frame of the video. The sound should be reproduced from background music, and combined to create a CT.

3   Results

3.1 Selection of Videos

Figure 2 and Table 1 show the results of the video impression survey. Except for monochrome videos, the Z-scores are in descending order as Excellent 1, 2, 3, Normal 1, 2, 3, and Poor 1, 2, 3.

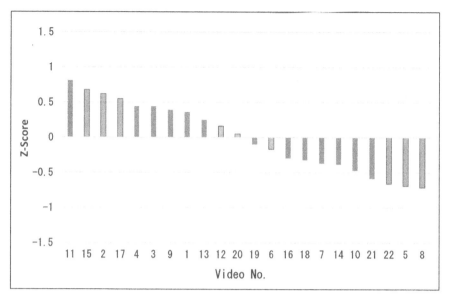

Fig. 2. Videos used to measure brain function

Table 1. Video and Z-score data for brain function

Group	Video no.	Z-score
Excellent	15	0.678
	2	0.619
	17	0.549
Normal	12	0.161
	20	0.051
	6	−0.168
Poor	22	−0.658
	5	−0.690
	8	−0.715

Figure 3 and Table 2 shows a comparison of the results of the brain function measurement and image selection. Excellent 2 had a Z-score of 0.619 in the selected experiment, while Excellent 2 had a Z-score of 0.356 in the first viewing and 0.23 in the second viewing. Excellent 2 was excluded because it was lower than the Normal group. The p-value of the 1st and 2nd views was 0.99, showing no significant difference.

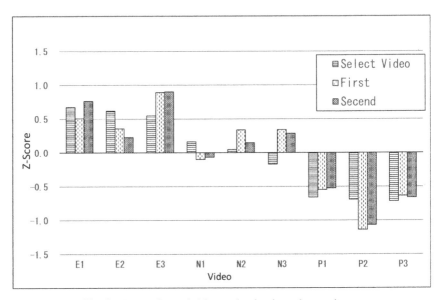

Fig. 3. Comparison of video evaluation in each experiment

Table 2. Z-Score for video evaluation in each experiment

Group	No.	Select	First	Second
Excellent	1	0.678	0.510	0.764
	2	0.619	0.356	0.230
	3	0.549	0.893	0.903
Normal	1	0.161	−0.096	−0.063
	2	0.051	0.336	0.146
	3	−0.168	0.341	0.286
Poor	1	−0.658	−0.543	−0.523
	2	−0.690	−1.137	−1.063
	3	−0.715	−0.635	−0.657

3.2 Measurement of Brain Activity

Table 3 shows the channel numbers, p-values, t-values, and the Broadmann area where the activity was assumed to have changed based on the t-tests: significant differences between the time series data of oxy-Hb when watching the television-commercial videos and the time series data when watching the CT. In the Broadmann area, R- indicates the right side, and L- indicates the left side. There was no significant difference in the DLPFC among the three groups in the second viewing (Table 3). Therefore, in the video evaluation, the DLPFC changed its activity only in the first viewing.

Table 3. Significant differences in brain activity between the control and CT images

Count	Group	Ch.	Place	p-value	t-value	Count	Group	Ch.	Place	p-value	t-value
		28Ch.	L-DLPFC	0.028	-2.353						
		25Ch.	R-FrP	0.044	-2.117						
	Excellent	19Ch.	L-Broca	0.039	-2.170		Excellent		No significant place		
		21Ch.	—	0.042	-2.166						
		44Ch.	—	0.049	2.060						
First		48Ch.	—	0.030	-2.338	Second					
								2Ch.	—	0.019	-2.641
	Normal	25Ch.	R-DLPFC	0.046	2.097		Normal	13Ch.	R-Broca	0.026	-2.483
								41Ch.	—	0.017	-2.483
		25,35Ch.	R-DLPFC	0.030	-2.929			2Ch.	—	0.037	-2.199
	Poor	26Ch.	R-FrP	0.045	-2.112		Poor	49Ch.	L-FrP	0.045	2.131
		9Ch.	—	0.044	-2.118						

Table 4 shows the significant difference between the first and second T t-tests of the cerebral function data in the number of views of the video, and the Broadmann area where changes in activity are assumed based on the p-values, t-values, and channel numbers.

Table 4. Significant difference in channels between the first and second viewings

Group	Channel	Brain	p-value	t-value
Excellent	28Ch.	L-DLPFC	0.021	-2.485
Normal	39Ch.	L-DLPFC	0.039	-2.174
	13Ch.	R-Broca	0.007	2.916
Poor	25Ch.	R-DLPFC	0.023	-2.335

The location of the preferred area and the time series changes of oxy-Hb are shown below. It is assumed that there was a significant difference in the colored Ch. in the figure of the preferred area. The red part of the graph of time series change was the part presenting the image. The graphs were plotted for each subject. The second viewing was omitted because there was no significant difference.

Excellent Group. oxy-Hb decreased overall at the first viewing (Fig. 5), and there was little variation among subjects. On the other hand, in the second viewing, there was significant variation among subjects (Fig. 6).

Normal Group. In the first viewing, oxy-Hb decreased overall (Fig. 8). No significant trend was observed in the second viewing (Fig. 10).

Poor Group. In the first viewing, the variation among participants was smaller than the second viewing (Figs. 12, 14). There was no significant difference in the location of Ch (Fig. 13).

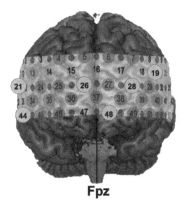

Fpz

Fig. 4. Ch. Location, a significant difference in the first viewing (Excellent)

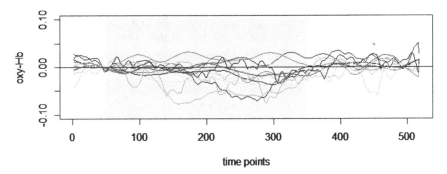

Fig. 5. Time series change of oxy-Hb in the first viewing (Excellent)

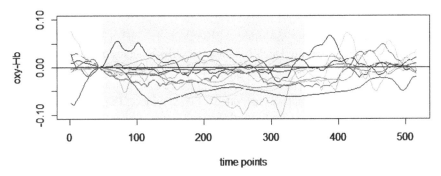

Fig. 6. Time series change of oxy-Hb in the second viewing (Excellent)

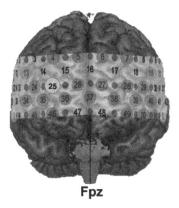

Fig. 7. Ch. location, a significant difference in the first viewing (Normal)

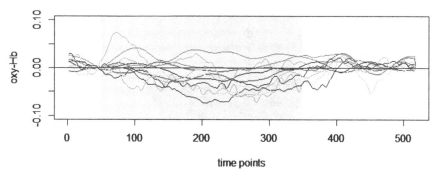

Fig. 8. Time series change of oxy-Hb in the first viewing (Normal)

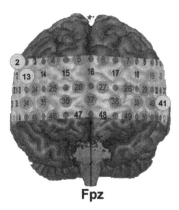

Fig. 9. Ch. Location, a significant difference in the second viewing (Normal)

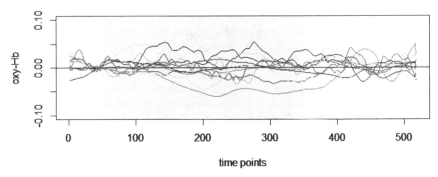

Fig. 10. Time series change of oxy-Hb in the second viewing (Normal)

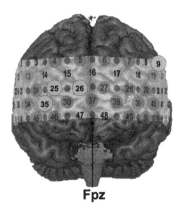

Fig. 11. Ch. Location, a significant difference in the first viewing (Poor)

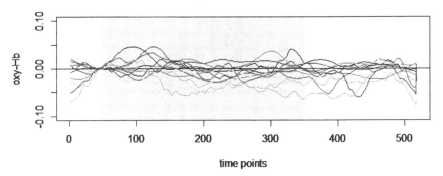

Fig. 12. Time series change of oxy-Hb in the first viewing (Poor)

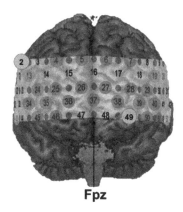

Fpz

Fig. 13. Ch. Location, a significant difference in the second viewing (Poor)

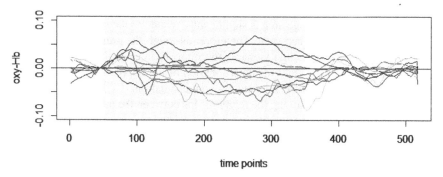

time points

Fig. 14. Time series change of oxy-Hb in the second viewing (Poor)

4 Discussion

The DLPFC was more active in the first-time viewing than in the second time viewing (Table 4). This suggests a psychological process in which the participants memorized the contents of the video when they watched it for the first time and drew it out again when they watched it for the second time. There may have been a change in the memory region deep in the brain.

In the time series change of oxy-Hb in the first-time viewing, the individual difference was larger than in the second viewing (Figs. 4, 5, 6, 7, 8, 9, 10, 11, 12). In the first-time viewing, all subjects followed the same psychological process. However, since the second time viewing was related to both changes in the prefrontal area and the memory area, the individual difference was large. What is recalled differs from person to person upon second time viewing. Therefore, it could be expected that the degree to which memories and impressions are connected may be different. In addition, the degree of change in oxy-Hb decreased for all participants in the first episode of the Excellent and Normal groups (Figs. 5, 8). The Excellent or Normal videos may have reduced not only the DLPFC but also the total oxy-Hb changes.

In this paper, the change in the DLPFC has been thought of as a feature of brain activity, which decides the first impression. The results of studies by Azehara and Kato [4] and Brent et al. [6] have indicated that the DLPFC activity decreases when the first impression is good. In addition, in this study, the activity of the DLPFC in watching the Excellent Z-score group decreased. Therefore, it can be said that the activity of the DLPFC decreased for the Excellent first impression. However, the reason the DLPFC activity decreased has not been clarified in any studies, including this one. Therefore, further clarification of this will lead to a better understanding of the relationship between impression determination and brain activity.

5   Conclusions

In this study, we analyzed the relationship between the change in impression and prefrontal area with the number of video views. As a result, it was proven that the activity of the DLPFC decreased when videos of which the first impression was 'good' were viewed. In addition, the storage area is significantly related to the second viewing, i.e., the video that has already been seen. Therefore, although the prefrontal area activity decreased compared with the first viewing, the impression evaluation did not change. Even if the video as an input and the determination of the impression as an output are the same, the brain activity differs depending on the situation.

This study revealed that among the relationships between the process of impression determination and frontal lobe activity, memory is related to impression determination. In this research, we analyzed the relationship between the cognitive process and the prefrontal cortex before deciding the impression.

NIRS is an objective technique that uses brain activity to estimate impressions perceived from various things or matters, and it enables us to know other people's feelings and impressions without verbalization. Consequently, optical topography can be an effective indicator of impression estimation in neuromarketing. optical topography is also expected to be utilized for brain-machine interfaces. Research issues being dealt with in this paper are regarded as one of the research themes of *kansei* engineering. Such an engineering technique to estimate impression using brain activity will be useful in a society that must accept differences in cognitive diversity.

Acknowledgement. We would like to thank Dr. Atsushi Maki (Hitachi Ltd.) for useful advice and discussions. This work was partially supported by JSPS KAKENHI Grant Numbers 25240043.

References

1. Picard, R.W.: Affective Computing. MIT Press, Cambridge (1997)
2. Lee, N., Brandes, L., Chamberlain, L., Senior, C.: This is your brain on neuromarketing: reflections on a decade of research. J. Market. Manag. **33**, 878–892 (2017). https://doi.org/10.1080/0267257X.2017.1327249
3. Hashimoto, Y., Tsuzuki, T.: Mismatch between interpretation of intention and results in multi-attribute decision-making: the choice blindness paradigm and decision-making by bounded rationality. Rikkyo Psychol. Res. **55**, 45–53 (2013)

4. Azehara, S., Nakata, T., Kato, T.: Analysis of relationship between impression of video and memory using fNIRS. In: Baldwin, C. (ed.) AHFE 2017. AISC, vol. 586, pp. 157–165. Springer, Cham (2018). https://doi.org/10.1007/978-3-319-60642-2_14

5. Tanida, H., Kato, T.: Verification of brain activity when watching TV commercials using optical topography. In: Fukuda, S. (ed.) AHFE 2018. AISC, vol. 774, pp. 76–81. Springer, Cham (2019). https://doi.org/10.1007/978-3-319-94944-4_9

6. Hughes, B.L., Zaki, J., Ambady, N.: Motivation alters impression formation and related neural systems. Soc. Cogn. Affect. Neurosci. 12(1), 49–60 (2017). https://doi.org/10.1093/scan/nsw147

7. McClure, S.M., Li, J., Tomlin, D., Cypert, K.S., Montague, L.M., Montague, P.R.: Neural correlates of behavioral preference for culturally familiar drinks. Neuron 44, 379–387 (2004). https://doi.org/10.1016/j.neuron.2004.09.019

8. Johansson, P., Hall, L., Sikstrom, S., Olsson, A.: Failure to detect mismatches between intention and outcome in a simple decision task. Science 310, 116–119 (2005). https://doi.org/10.1126/science.1111709

9. Villringer, A., Planck, J., Hock, C., Schleinkofer, L., Dirnagl, U.: Near infrared spectroscopy (NIRS): a new tool to study hemodynamic changes during activation of brain function in human adults. Neurosci. Lett. 154, 101–104 (1993). https://doi.org/10.1016/0304-3940(93)90181-j

10. Muto, T., Tsumori, S.: Mou ichido mitai nihon no CM 50 nenn (Japanese TV commercial in 50 years which we want to watch once again). Avex Entertainment Inc. (2010)

Visualizing Emotion and Absorption Through a Low Resolution LED Array:
From Electroencephalography to Internet of Things

Xiaobo Ke$^{(\boxtimes)}$ ⓘ and Christian Wagner ⓘ

School of Creative Media, City University of Hong Kong, Hong Kong, China
xiaoboke-c@my.cityu.edu.hk, c.wagner@cityu.edu.hk

Abstract. Electroencephalography (EEG) is a popular method for detecting and recording electrical activity in the human brain. Using this method, we can explore the relationships between specific brain activities and certain mental states. Even though neuroscience measurements are reliable and useful, the specific methods are not quite popular within human-computer interaction (HCI) research and practice, since traditional devices for neurophysiological data acquisition are professional knowledge dependent and expensive. The emergence of inexpensive, off-the-shelf wireless EEG devices encourages researchers and designers to probe the new perspective of HCI. In this paper, we present a pilot study which attempts to connect the EEG data obtained by a wireless EEG device (i.e., Emotiv EPOC plus) to an electronic device (i.e., a Particle-based 8*8*8 LED array) enabled by the infrastructure of the Internet of Things (IoT). Specifically, this study visualizes the two general types of human emotions and the human absorptive state on the LED array through multidimensional display patterns. We hope this study can be regarded as the pilot for the brain-computer interface building with commercial EEG products and electronic devices enabled by IoT. Furthermore, the study can also serve as reference for creative scientific visualization of human mental states with physical electronic devices.

Keywords: Emotion · Absorption · Visualization · Electroencephalography (EEG) · Internet of Things (IoT)

1 Introduction

Electroencephalography (EEG) is a method that detects and records electrical activity in the human brain using electrodes attached to the scalp [13]. EEG is a popular test for brain activity analysis [11, 44]. However, the traditional devices for the measurement of EEG signal are usually specialized knowledge required, expensive, and unwieldy. In this case, despite the high accuracy and temporal resolution, these medical level devices are not much popular in the practice and research on human-computer interaction (HCI). Recently, the emergence of inexpensive wireless EEG devices results in augmented flexibility and mobility over traditional devices [30]. The Emotiv (www.emotiv.com), NeuroSky (store.neurosky.com), and Muse (choosemuse.com) are the representatives of

© Springer Nature Switzerland AG 2020
D. D. Schmorrow and C. M. Fidopiastis (Eds.): HCII 2020, LNAI 12196, pp. 142–156, 2020.
https://doi.org/10.1007/978-3-030-50353-6_11

this product line. These innovative and user-friendly devices have also opened the new research perspective for research methods and design ideas in the field of HCI [27].

Within the research field of HCI, scholars are usually interested in the users' emotional experience (say, the enjoyment and excitement) (e.g., [24, 25]) and the immersive experience (say, the flow experience and absorption) (e.g., [43]). The overall attraction of the hedonic information systems, especially video games, is highly dependent on the emotional and immersive feeling which users can experience during the system engagement [43]. To understand the user experience, the common assessment mainly relies on the survey performed after the engagement with systems (e.g., [24, 43]). This measuring strategy has two weaknesses: (1) the threat to feeling distortion in the self-report survey [3] and (2) lacking a dynamic view of the interaction process [29]. In this case, the real-time neurophysiological measurement (say, EEG) is a promising data collection idea that can overcome the weaknesses of the post-engagement survey. Recently, some high-quality studies on user experience start to collect the neurophysiological data to examine and support their arguments (e.g., [31]).

Besides, these wireless EEG devices (say, Emotiv) also contribute to a promising research and design stream which is called the brain-computer interface (BCI). A BCI is defined as a communicational channel (independent from the brain's normal output pathways) connecting the brain to other electronic devices [46]. Relevant literature body is expanding to help the development of BCI. For example, emotion recognition, focused thought detection, and specific brain signal extraction through EEG devices are the active research areas (e.g., [18, 22, 38]). These studies are the signal extraction and understanding for "reading the brain". However, the literature efforts towards the connection and integration between electronic devices and the discussed wireless neuroscience devices are limited [27, 47]. For the physical integration of various objects and their unique capabilities within a system, the Internet of Things (IoT) is a helpful and emerging solution [40]. IoT is an ecosystem for the interrelation of objects, creatures and/or human which are offered unique sign and the communication ability over a network with self-configuring capability [45]. For the development of IoT, Particle (www.particle.io), Raspberry Pi (www.raspberrypi.org), and Arduino (www.arduino.cc) are some of well-known IoT-ready hardware devices and communities.

Thus, this study aims to present a basic BCI system that connects the EEG data to an electronic device enabled by IoT. Given the strength of EEG devices is the reflection of the human emotional and absorptive experience in real-time, this study specifically attempts to visualization the real-time human emotional and absorptive experience on a Particle-based LED array. Overall, this study can be regarded as the pilot study for the research on the BCI system building with commercial EEG devices under the basic framework of IoT. In addition, this study can also be the pilot for artistic and scientific visualization of various human mental states on the different physical electronic devices.

2 Literature Review

2.1 Classification Framework for Affective States

According to the definition provided by American Psychological Association (APA), emotion refers to a complex reaction pattern, involving experiential, behavioral, and physiological elements [2]. Therefore, emotion describes a complicate psycho-neural process of interaction between subjective and objective variables mediated by neural and hormonal systems [26, 37]. For the improvement of systems design and user experience, the fulfillment of users' emotional experience is one of the important dimensions [25].

Based on literature review, this paper summarized general frameworks for the human emotions' identification and label. As shown in Table 1, literature shows three perspectives for scholars to understand the classification of human emotions.

Table 1. The summary of three classification frameworks for emotions.

Perspectives for emotion classification	Description of the frameworks	Research focus and research scenario	Supporting references
Discrete emotions	The summary of the specific types of emotions	This perspective is usually the basic research background for research on emotion and human facial expression	[14–16, 42]
Fusion view of emotions	This perspective is interested in the relationship between the specific types of emotions	This perspective is applicable to help the people understand the generation of humans' complicated emotional states	[4, 9, 39]
Dimensional axes of emotions	Specific emotions can be located in the specific place within the multidimensional coordinate	This perspective is usually be used in the affective and emotional computing to recognize human emotions algorithmically and technically	[32, 35, 41]

Firstly, one research stream is the perspective of discrete emotions. For example, serial research (e.g., [14–16]) helps us to understand the relationship of basic types of human emotions via facial expression and cultural difference. For more specific and complex emotions, a tree-structure framework of emotion states was described by Shaver et al. [42], which include more specific types. The second research stream of the emotion classification is the fusion view of the emotions. For example, serial research (e.g., [4, 9, 39]) displays a wheel of eight emotions: joy, trust, fear, surprise, sadness, disgust, anger and anticipation, which can be dyadic or combined to generate the new type of emotions. The last but not least research stream investigates the dimensional axes of emotions. From this perspective, the fundamental dimensions for the emotion classification are the valance dimension (say, pleasure-displeasure states) and arousal dimension (say, the excitement level of the individual) [32]. The valance-arousal framework reflects the "core affect" of human emotions [41]. This framework is also usually the theoretical support for EEG-based emotion recognition (e.g., [35]).

Drawn on the valance-arousal framework, this research selected the interesting (a human emotion) to represent the valance level and the excitement (a human emotion) to reflect the arousal level of the individual. In addition, this research also included the focus level to show the absorptive experience of individuals.

2.2 Neurophysiological Measurements and Analysis Perspective

Brain activities can be monitored and tracked using various neurophysiological measurements, such as standard scalp-recording electroencephalogram (EEG), magnetoencephalogram (MEG), functional magnetic resonance imaging (fMRI), electrocorticogram (ECoG), near-infrared spectroscopy (NIRS), and so forth [1].

In the research area of neuroscience, the identification of the relationship between the human brain activities and the individual mental states is of main research interest to researchers. The emotional state identification is one of the most popular and promising research topics in neuroscience and other relevant fields (say, affective computing). Furthermore, the studies usually take advantage of the EEG data collected from EEG scalp recordings to research the relationship of brain activities with emotional states (e.g., [36]). This implies the potential of using the EEG-based signal to reflect the specific type of emotional states.

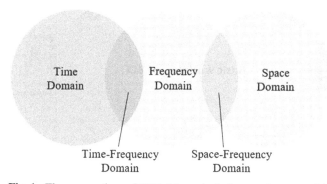

Fig. 1. The perspectives of EEG data analysis for emotion recognition

Specifically, Fig. 1 displays analytical perspectives that can help researchers understand emotions through the indicators extracted from EEG data. From the time domain, the common EEG properties that can be used in this domain are the event-related potential [21], signal power [23], and Hjorth indicators [20]. From the frequency domain, the first step for the data analysis is data transformation from the time domain to the frequency domain using Fourier transfer. Common EEG properties that can be used to recognize emotions are features of the signal power [23] and difference entropy [48]. The domain of time-frequency includes the time window in the Fourier transfer which can increase the ability for processing the unstable signal [6]. As for the space domain, the location information of the electrodes is included in the analysis. This domain usually combines with the information obtained from the frequency or time-frequency to help the researchers classify the emotional states of the human (e.g., [7]).

2.3 Brain-Computer Interface (BCI)

Brain-computer interface (BCI) (also called the mind-machine interface or brain-machine interface) is an integrated platform enabling the direct communication and interaction between the human brain and peripheral devices. The more classical and clear definition of BCI is: A BCI is a communicational channel that connects the brain to a computer or other electronic devices [10]. BCI is also an electrode-computer construct that extracts and decodes information from the nervous system to create functional outputs [28]. The explicit difference between BCI and the traditional HCI paradigm is the dependency on the brain's normal output pathways of peripheral nerves, muscles, and skins [46]. The common focus of current advancing works of HCI research and design is still mainly on the human peripheral nerves, muscles, and skins (e.g., [8]). However, the BCI aims to establish the direct communicational channels (say, bypassing the motor system) between the brain and electronic peripherals. In other words, the BCI is an effective way to augment human cognition in the interaction with electronic devices and facilitate neural plasticity and motor learning for medical rehabilitation [28].

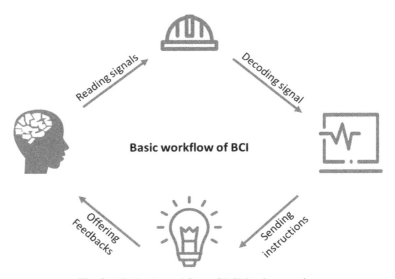

Fig. 2. The basic workflow of BCI implementation

The basic workflow for the implementation of a BCI system is shown in Fig. 2. There are four fundamental steps to realize the interaction between human brain and the electronic devices [28]: (1) Reading the raw signals from the human brain; (2) processing the signals and understanding the meanings of signals; (3) mapping the signals to the functional outputs and sending the instructions to electronic devices; (4) receiving feedbacks from the performance of electronic devices. According to the workflow, we can see that the starting point of the BCI is measuring neurophysiological signals from the human brain. Although several measurements and data can be obtained from brain activities, the most common and optimal choice for BCI implementation is still the EEG [10]. Thus, this research also follows this selection.

3   Design and Materials

3.1   Design Idea

The fundamental idea of this project is to stream the real-time EEG data to an electronic device enabled by IoT. Specifically, this project attempts to visualize the human's emotional states and absorption by using various lighting and display patterns of a low-resolution LED array.

In this project, two specific emotional states and the absorptive state are planned to visualize on the LED array: the positive affect (i.e., interesting), the excitement, and the focus. Accordingly, three patterns or dimensions of LED array display are designed. In the project, a dynamic and luminous sphere shell is displayed in an 8*8*8 LED array. The color of the shell indicates the positive affect of the individual; the dynamic changing speed of the shell indicates the excitement of the individual; the maximum size of the shell indicates the focus level of the individual.

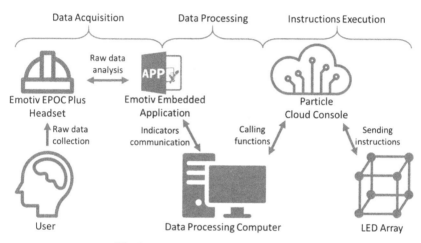

Fig. 3. The infrastructure of the systems

To realize this idea, we need to build up a system including hardware and software implementation and communication. The abridged general view of the whole system is shown in Fig. 3. For the EEG data reading and processing, we applied the Emotiv headset and its embedded application. For the EEG data communication to the computer, we adopted the WebSockets solution. As for connecting and sending the functional instructions to the LED array, we rely on the Particle's could application program interfaces (APIs). The specific introduction of the hardware and software implementation is listed in the following two subsections.

3.2   Hardware Implementation

According to the basic workflow of BCI implementation, one critical hardware we need would be a device helping us measure the brain activities and transfer the data of these actives to the computer. In addition, another critical hardware would be the electronic device that responds to the instructions sent by the computer.

In this research, the device for measuring the EEG data from the human brain is the EMOTIV EPOC plus (See Fig. 4). EPOC plus is a portable and wireless EEG device with 14 channels designed for scalable and contextual human brain research and advanced brain-computer interface applications (for more details, please refer to www.emotiv.com/epoc). Even though the EPOC plus only has 14 channel which is less than the traditional medical use EEG devices with 32, 128, or 256 channels, the data reliability and validity as well as the measurement stability are still strong enough for the researchers to identify the pattern of brain activities [5]. In addition, EPOC plus offers ample APIs which provides the possibility for us to connect the EEG data to the different electronic devices. This is a critical part of the communication between the human brain and devices. Given the reliability measurement and the tools (say, APIs) supporting secondary development, this research selected the EPOC plus as the device for EEG data collection.

Fig. 4. EMOTIV EPOC plus (the EEG device used in this study)

The main aim of this project is to connect the EEG data to the electronic device enabled by IoT. To achieve this goal, the electronic device for responding the instructions is an 8*8*8 LED array (see Fig. 5).

Fig. 5. An 8*8*8 LED array powered by Particle firmware

The control center for this LED array is a firmware called Photon produced by Particle. Photon is a tiny Wi-Fi IoT device for creating connected projects and products. In addition, Particle (for more information, please refer to www.particle.io) is a fully all-in-one platform that provides the integrated development environment including hardware kits, software support, and connectivity solution. This would be an easier way for the researcher and designers who do not have a strong background in engineering and computer science to develop their physical prototype.

3.3 Software Implementation and Data Acquisition

For the part of data acquisition, we used an application provided by Emotiv to obtain the data extracted by the EPOC plus. Based on their detection and classification algorithm as well as their previous experiments, the Emotiv is capable to provide the numerical indicators of valance and arousal which reflect the emotional states of individuals from the perspective of dimensional axes of emotion (for this perspective, please refer to Sect. 2.1 Classification Framework for Affective States). Emotiv's application helps us obtain these indicators representing the individual's emotional status. In addition, we also included an indicator measuring individual's focus level in tasks, which reflects user's concentration experience within system engagement.

For the LED array programming, the device's operation system of Particle features an easy-to-use programming framework to help the designers and researchers develop applications to run on the devices (Photon in this research). The programing platform for the Photon is an integrated development environment (IDE), Particle provides the Web version and Desktop version of IDEs which enable the designer and researcher to review and revise the programs anywhere and anytime.

For the data communication from EPOC plus to the LED array, the WebSockets solution is adopted. For the data communication part, we used Python to set up the communication channel. The Websockets relies on the library of Websocket (please refer to pypi.org/project/websockets). The Websockets thread is responsible to keep the thread alive and always connected to the server.

4 Showcase of the Prototype

Figure 6 shows the whole basic systems built for this research. The Emotiv data acquisition component was built upon the APIs provided on ozancaglayan/python-emotiv3. These components connect to the corresponding USB port for initializing and establishing the communication with the headset of EPOC plus. From the headset, we obtain the indicators reflecting the individual's positive affect (i.e., interesting level), arousal (i.e., excitement level), and the absorption (i.e., focus level). As far as the headset remains connected, every newly received packet is transmitted to the local server for the data translation to the LED array by invoking the Websockets component. According to the real-time EEG data received, the local server will call the functions configured on the could console of the LED array (enabled by the Photon @ Particle). Thus, the LED array can show the specific pattern to visualize the human emotional states and the absorptive level.

Fig. 6. The system built by this study

When the system is on, the LED array will display a dynamic luminous shell changing the size repeatedly. Figure 7 shows the initial status of the prototype, namely the starting status before the LED array receives any instruction evoked by the Emotiv device.

Fig. 7. Initial status of the LED array in the system

Figure 8 shows that the colors of the shell indicate individuals' positive affect level. Blue indicates the individual with low positive affect, purple means the middle level and red implies the high level.

Fig. 8. Color patterns for indicating the positive affect level (Color figure online)

In the prototype, the size of the shell is dynamic and tautologically changing. Thus, the changing speed represents the arousal level of the individual. Specifically, the faster the changing rate is, the higher the arousal level of the individual is. In addition, the maximum size of the shell represents the absorption of the individual. As shown in Fig. 9, the different maximum sizes of the shell imply the different levels of an individual's absorption (i.e., focus level). The bigger the maximum size is, the higher the focus level of the individual is.

Fig. 9. Maximum size of the shell for indicating the focus level (Color figure online)

A showcase of the prototype is shown in Fig. 10. In this showcase, a gamer who wears the Emotiv EPOC plus is playing a simulated racing game.

In the previous literature, scholars found that the simulation racing game can trigger the player's positive affect and arousal level (e.g., [17]). Furthermore, the positive game experience could also lead the gamer to immerse in the game engagement which makes the gamer get a high focus level (e.g., [33]). Thus, it is suitable for us to use the simulation racing game to evoke gamer's relevant emotional states and absorption which are of the interest to this project. Specifically, we use the prototype of this project to show this gamer's positive affect, arousal, and the focus level while the gamer is playing the simulation racing game. The results of this showcase reflect the satisfactory quality of the system.

Fig. 10. A showcase of the prototype system

5 Discussion

In this research, we developed a basic BCI system for visualizing human emotions and absorption. In this section, we would like to discuss the suitable scenarios in which the system is helpful for their user. In addition, this section also discusses limitations and the plan for the further improvement of the system.

5.1 Applicable Scenarios

The core function of the system built in this study is the visualization of the human mental states (in the study, we visualize the two general types of emotions and the absorption). Therefore, the applicable scenarios for this system would be the contexts in which the human mental states are focal for the users. In this subsection, we discuss a specific context that may be suitable for people to use the system of this study. This specific context is the training of eSports players.

ESports is a popular spectator sport and leisure activity for electronic game players and even non-gamers. The huge number of eSports spectators is one of the critical drivers for the spectacular growth of the eSports industry [19]. To help the sustainable development of the eSports market, the high-quality performance among the eSports players is crucial for eSports tournaments to continue to attract a huge number of the audience.

The training of eSports players is the major measure to improve their performance. Within the training, emotion management is also the major concern of coaches and players. Numerous studies on sports science and sports psychology indicates that inappropriate emotional states usually negatively influence athletes' in-game performance. For example, the athletes' performance will become chocking under extreme pressure. Chocking refers to "an acute and considerable decrease in skill execution and performance when self-expected standards are normally achievable, which is the result of increased anxiety under perceived pressure" [34]. On the contrary, some studies shed light on the influence of positive emotional states (say, enjoyment) on the component of performance excellence [12]. These indicate that the emotional factor is also the critical success factor for the high-quality performance of the eSports players. Figure 11 display the contextualized use of the system in training eSports players.

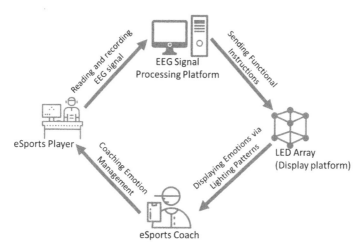

Fig. 11. The contextual use of the system in eSports training

Figure 11 indicates that our systems as the emotion and absorption visualization platform can help the coaches and the eSports players monitor their emotional changes and focus level. Besides, this monitor also can be regarded as real-time feedback which can help these professional players to adjust their emotional state and improve their concentration during the gaming. The contextual aim of the system in the training is to be the assistance helping eSports players understand and adjust their emotional state to the optimal status when they are in gaming.

Apart from the specific scenarios discussed above, there are some other potential contexts that are applicable to this system. For pragmatic use, this system is also a useful tool for the people who have difficulty in recognizing other's emotions through common expressions approaches (say, facial expressions or other non-verbal expressions). This system can be regarded as an explicit expression assistance to help these people understand emotions expressed by others. For artistic or hedonic use, this system is also a potential tool in media art or emotional game development. For example, well-known media artist Maurice Benayoun[1] has pioneered an acclaimed media art project named "The Brain Factory"[2] which is also based on EEG data. This implies that the artistic visualization of the brain activates could also be the potential context of this system.

5.2 Limitations and Further Improvement

This research is a pilot study of the BCI system for mental state visualization and control. As a pilot study of the whole project, this study has several limitations that offer

---

[1] http://www.scm.cityu.edu.hk/profile/M.BENAYOUN.
[2] https://benayoun.com/moben/2017/07/04/the-brain-factory-trailer/.

the opportunity for future improvement. First, the visualization patterns for the human mental state are simple and the types of human mental states visualized on the LED array are limited. Thus, the future effort can be spent on the fancier display pattern development and the visualization of more human mental states. Plus, the interactivity of this BCI is limited. In this study, we only develop the basic interface for the display of the mental state. Thus, the actual interaction design should be considered for further improvement. In addition, this system only shows the overview of mental states. However, the location information of the brain is ignored. Hence, future improvement can also be the display of the brain's area location information on the LED devices.

6   Conclusion

This study built a system that connects EEG data to a low resolution LED array enabled by IoT. The system visualizes human positive affect through the color changes of the LED; arousal through the changing speed of the dynamic shell shown on the LED array; and focus level through the maximum size of the shell. This study as the pilot research can be regarded as the reference for researchers, artists, and designers who plan to connect the neurophysiological signals to the commercial off-the-shelf products for scientific or creative purposes.

Acknowledgement. The research presented in this article was supported in part by the Research Grants Council of the Hong Kong SAR under GRF Project No. 11610419.

References

1. Amiri, S., Fazel-Rezai, R., Asadpour, V.: A review of hybrid brain-computer interface systems. Adv. Hum.-Comput. Interact. **2013**(187024), 1–8 (2013)
2. APA Dictionary of Psychology. https://dictionary.apa.org/emotion. Accessed 19 Jan 2020
3. Ariely, D., Berns, G.S.: Neuromarketing: the hope and hype of neuroimaging in business. Nat. Rev. Neurosci. **11**(4), 284–292 (2010)
4. Athar, A., Khan, M.S., Ahmed, K., Ahmed, A., Anwar, N.: A fuzzy inference system for synergy estimation of simultaneous emotion dynamics in agents. Int. J. Sci. Eng. Res. **2**(6), 35–41 (2011)
5. Badcock, N.A., Mousikou, P., Mahajan, Y., De Lissa, P., Thie, J., McArthur, G.: Validation of the emotiv EPOC® EEG gaming system for measuring research quality auditory ERPs. PeerJ. **1**(e38), 1–17 (2013)
6. Behnam, H., Sheikhani, A., Mohammadi, M.R., Noroozian, M., Golabi, P.: Analyses of EEG background activity in Autism disorders with fast Fourier transform and short time fourier measure. In: 2007 International Conference on Intelligent and Advanced Systems, pp. 1240–1244. IEEE, Kuala Lumpur (2007)
7. Blankertz, B., Tomioka, R., Lemm, S., Kawanabe, M., Muller, K.R.: Optimizing spatial filters for robust EEG single-trial analysis. IEEE Signal Process. Mag. **25**(1), 41–56 (2007)
8. Cai, S., Ke, P., Narumi, T., Zhu, K.: ThermAirGlove: a pneumatic glove for thermal perception and material identification in virtual reality. In: 27th IEEE Conference on Virtual Reality and 3D User Interfaces. IEEE, Atlanta (2020, in press)

9. Chafale, D., Pimpalkar, A.: Sentiment analysis on product reviews using Plutchik's wheel of emotions with fuzzy logic. ABHIYANTRIKI Int. J. Eng. Technol. **1**(2), 1–8 (2014)
10. Cheng, M., Gao, X., Gao, S., Xu, D.: Design and implementation of a brain-computer interface with high transfer rates. IEEE Trans. Biomed. Eng. **49**(10), 1181–1186 (2002)
11. Daliri, M.R.: Kernel earth mover's distance for EEG classification. Clin. EEG Neurosci. **44**(3), 182–187 (2013)
12. Derakshan, N., Eysenck, M.W.: Introduction to the special issue: emotional states, attention, and working memory. Cogn. Emot. **24**(2), 189–199 (2010)
13. Ding, Y., Cao, Y., Qu, Q., Duffy, V.G.: An exploratory study using Electroencephalography (EEG) to measure the smartphone user experience in the short term. Int. J. Hum.-Comput. Interact. **36**, 1008–1021 (2020). in press
14. Ekman, P.E.: Facial expression and emotion. Am. Psychol. **48**(4), 384–392 (1993)
15. Ekman, P.E., Davidson, R.J.: The Nature of Emotion: Fundamental Questions. Oxford University Press, Oxford (1994)
16. Ekman, P.E., Friesen, W.V.: Constants across cultures in the face and emotion. J. Personal. Soc. Psychol. **17**(2), 124–129 (1971)
17. Fischer, P., Kubitzki, J., Guter, S., Frey, D.: Virtual driving and risk taking: do racing games increase risk-taking cognitions, affect, and behaviors? J. Exp. Psychol.-Appl. **13**(1), 22–31 (2007)
18. Gao, X., Xu, D., Cheng, M., Gao, S.: A BCI-based environmental controller for the motion disabled. IEEE Trans. Neural Syst. Rehabil. Eng. **11**(2), 137–140 (2003)
19. Hamari, J., Sjöblom, M.: What is eSports and why do people watch it? Internet Res. **27**(2), 211–232 (2017)
20. Hjorth, B.: EEG analysis based on time domain properties. Electroencephalogr. Clin. Neurophysiol. **29**(3), 306–310 (1970)
21. Hruby, T., Marsalek, P.: Event-related potentials-the P3 wave. Acta Neurobiol. Exp. **63**(1), 55–63 (2002)
22. Jatupaiboon, N., Pan-ngum, S., Israsena, P.: Emotion classification using minimal EEG channels and frequency bands. In: 10th International Joint Conference on Computer Science and Software Engineering, pp. 21–24. IEEE, Piscataway (2010)
23. Jenke, R., Peer, A., Buss, M.: Feature extraction and selection for emotion recognition from EEG. IEEE Trans. Affect. Comput. **5**(3), 327–339 (2014)
24. Ke, X., Du, H.S., Wagner, C.: Encouraging Individuals to go green by gamification: an empirical study. In: 23rd Pacific Asia Conference on Information Systems, pp. 1–13. Association of Information Systems, Xi'an (2019)
25. Ke, X., Wagner, C.: The impact of game peripherals on the gamer experience and performance. In: Schmorrow, D.D., Fidopiastis, C.M. (eds.) HCII 2019. LNCS (LNAI), vol. 11580, pp. 256–272. Springer, Cham (2019). https://doi.org/10.1007/978-3-030-22419-6_18
26. Kleinginna Jr., P.R., Kleinginna, A.M.: A categorized list of emotion definitions, with suggestions for a consensual definition. Motiv. Emot. **5**, 345–379 (1981). https://doi.org/10.1007/BF00992553
27. Konstantinidis, E., Conci, N., Bamparopoulos, G., Sidiropoulos, E., De Natale, F., Bamidis, P.: Introducing Neuroberry, a platform for pervasive EEG signaling in the IoT domain. In: 5th EAI International Conference on Wireless Mobile Communication and Healthcare, pp. 166–169. Institute for Computer Sciences, Social-Informatics and Telecommunications Engineering, London (2015)
28. Krucoff, M.O., Rahimpour, S., Slutzky, M.W., Edgerton, V.R., Turner, D.A.: Enhancing nervous system recovery through neurobiologics, neural interface training, and neurorehabilitation. Front. Neurosci. **10**(584), 1–23 (2016)

29. Lallemand, C., Gronier, G., Koenig, V.: User experience: a concept without consensus? Exploring practitioners' perspectives through an international survey. Comput. Hum. Behav. **43**, 35–48 (2015)

30. Lau-Zhu, A., Lau, M.P.H., McLoughlin, G.: Mobile EEG in research on neurodevelopmental disorders: opportunities and challenges. Dev. Cogn. Neurosci. **36**(100635), 1–14 (2019)

31. Li, M., Jiang, Q., Tan, C.H., Wei, K.K.: Enhancing user-game engagement through software gaming elements. J. Manag. Inf. Syst. **30**(4), 115–150 (2014)

32. Mauss, I.B., Robinson, M.D.: Measures of emotion: a review. Cogn. Emot. **23**(2), 209–237 (2009)

33. McGloin, R., Farrar, K.M., Fishlock, J.: Triple whammy! Violent games and violent controllers: Investigating the use of realistic gun controllers on perceptions of realism, immersion, and outcome aggression. J. Commun. **65**(2), 280–299 (2015)

34. Mesagno, C., Hill, D.M.: Definition of choking in sport: re-conceptualization and debate. Int. J. Sport Psychol. **44**(4), 267–277 (2013)

35. Nie, D., Wang, X.W., Shi, L.C., Lu, B.L.: EEG-based emotion recognition during watching movies. In: 5th International IEEE/EMBS Conference on Neural Engineering, pp. 667–670. IEEE, Cancun (2011)

36. Onton, J.A., Makeig, S.: High-frequency broadband modulation of electroencephalographic spectra. Front. Hum. Neurosci. **3**(61), 1–18 (2009)

37. Panksepp, J.: Affective Neuroscience: The Foundations of Human and Animal Emotions. Oxford University Press, Oxford (1998)

38. Pham, T.D., Tran, D.: Emotion recognition using the emotiv EPOC device. In: Huang, T., Zeng, Z., Li, C., Leung, C.S. (eds.) ICONIP 2012. LNCS, vol. 7667, pp. 394–399. Springer, Heidelberg (2012). https://doi.org/10.1007/978-3-642-34500-5_47

39. Plutchik, R.: The nature of emotions: human emotions have deep evolutionary roots, a fact that may explain their complexity and provide tools for clinical practice. Am. Psychol. **89**(4), 344–350 (2001)

40. Porter, M.E., Heppelmann, J.E.: How smart, connected products are transforming competition. Harv. Bus. Rev. **92**(11), 64–88 (2014)

41. Russell, J.A., Barrett, L.F.: Core affect, prototypical emotional episodes, and other things called emotion: dissecting the elephant. J. Personal. Soc. Psychol. **76**(5), 805–819 (1999)

42. Shaver, P., Schwartz, J., Kirson, D., O'connor, C.: Emotion knowledge: further exploration of a prototype approach. J. Personal. Soc. Psychol. **52**(6), 1061–1086 (1987)

43. Suh, A., Cheung, C.M., Ahuja, M., Wagner, C.: Gamification in the workplace: the central role of the aesthetic experience. J. Manag. Inf. Syst. **34**(1), 268–305 (2017)

44. Taghizadeh-Sarabi, M., Daliri, M.R., Niksirat, K.S.: Decoding objects of basic categories from electroencephalographic signals using wavelet transform and support vector machines. Brain Topogr. **28**(1), 33–46 (2015)

45. Yang, A.M., Yang, X.L., Chang, J.C., Bai, B., Kong, F.B., Ran, Q.B.: Research on a fusion scheme of cellular network and wireless sensor for cyber physical social systems. IEEE Access **6**, 18786–18794 (2018)

46. Wolpaw, J.R., Birbaumer, N., McFarland, D.J., Pfurtscheller, G., Vaughan, T.M.: Brain–computer interfaces for communication and control. Clin. Neurophysiol. **113**(6), 767–791 (2002)

47. Zaki, M., Alquraini, A., Sheltami, T.: Home automation using EMOTIV: controlling TV by brainwaves. J. Ubiquitous Syst. Pervasive Netw. **10**(1), 27–32 (2018)

48. Zheng, W.L., Lu, B.L.: Investigating critical frequency bands and channels for EEG-based emotion recognition with deep neural networks. IEEE Trans. Auton. Ment. Dev. **7**, 162–175 (2015)

Methodology for Detection of ERD/ERS EEG Patterns Produced by Cut Events in Film Fragments

Javier Sanz Aznar[1]([✉]) [ID], Carlos Aguilar-Paredes[1] [ID], Lydia Sánchez-Gómez[1] [ID],
Luis Emilio Bruni[2] [ID], and Andreas Wulff-Abramsson[2] [ID]

[1] Universitat de Barcelona, Barcelona, Spain
javier.sanz@ub.edu
[2] Aalborg University, Copenhagen, Denmark

Abstract. The goal of this communication is to create a framework to isolate the neural reactions registered through EEG as a consequence of a specific input, among all those caused by an audiovisual. In order to do that, we analysed the neuronal register of the power change reactions related to specific cinematographic techniques, in this case *the shot change by cut*.

To research the shot change by cut through the neuronal record, one could use *ad hoc* audiovisual material specifically made for the experiment were the inputs are artificially introduced, or one could work with commercial cinematographic material, in order to have a more ecological approach For the latter approach, a more complex signal analysis process is needed because in the neuronal register the cut reactions are difficult to isolate.

For this experiment we used the EEG records of 21 subjects watching fragments extracted from 4 feature films that represent different styles and technical approaches. From the 21 user's records we created a model signal for each film and compared the power change between the different model signals of each films through permutations test, Spearman correlation and analysis of slopes. Through an automated process with sliding time windows, we were able to locate those temporal lapses, electrodes and frequency bands that show reactions to the shot change by cut in the power change that have synchronous reactions in all the model signals. We also located the neuronal reactions that suppose representative variations in the ERD/ERS due to the cut input.

Keywords: Methodology · EEG · Power change · ERD/ERS · Multiple inputs · Isolate inputs · Neurocinematics · Permutation test · Spearman · Slope analysis · Shot change · Audiovisual · Film

1 Introduction

To determine the neuronal behaviour elicited on the spectator by such an event as a shot change in any feature film presents more challenges than if one were to use audiovisual stimuli created *ad hoc* for isolated laboratory studies. However, being able to analyse

© Springer Nature Switzerland AG 2020
D. D. Schmorrow and C. M. Fidopiastis (Eds.): HCII 2020, LNAI 12196, pp. 157–175, 2020.
https://doi.org/10.1007/978-3-030-50353-6_12

real cases instead of cases designed for a particular study, allows us to get closer to the viewer's neuronal response to the reality of a film. This kind of research could open the doors to the introduction of new kinds of biofeedback in the emerging field of Adaptive Storytelling.

The cognitive study of shot changes in films has been approached from fMRI [1], ERD/ERS [2] and blinks [3] among others. Usually studies based on ERD/ERS use material specifically made for laboratory studies. The aim of this study is to analyse the ERD/ERS responses to cuts in real film fragments. To achieve this an adaptive procedure is developed to effectively locate areas of the ERD/ERS that coexist across cuts to avoid locating ERD/ERS areas produced by a stochastic kind of cut. To obtain this analysis, we needed to identify reactions to different types of cuts within different aesthetic contexts.

The development of brain-computer interfaces allows a new form of evolution in the cinematographic discipline, for example in the emerging field of adaptive storytelling and the interactive cinema [4–6], but in order to have a deep control of the interaction it is necessary to break down how the cinematographic techniques affect the cognitive system of the viewer. The first human-film iterations have already appeared where the viewer intervenes in the edition [7], but there is still a great need for evolution to achieve a full brain-computer iteration that determines the interaction with the film.

To accomplish the objective of locating patterns of neuronal reactions to the shot change event through the ERD/ERS, we developed a methodology that combines the permutation test and Spearman's correlation test on a six-sample size sliding time windows of the power change for every pair of model signals. When a specific time window for a specific frequency band and electrode appears correlated in all possible combinations of pairs, we can consider that we have located a neuronal reaction in ERD/ERS that is responding to the cuts.

2 Previous Researches

Among the previous studies on cinematographic technique through the ERD/ERS, it is worth mentioning those conducted by Heimann [2, 8] and Martín-Pascual [9], due to the similarities with the present investigation. Different from the present studies, they all created their own filmed material to develop their experiments. They try to replicate in the audiovisual fragments the same conditions in each clip introducing the minimum number of possible changes. So, in the work «Moving mirrors» [8] they explain how they filmed the same situation with different camera approach techniques, trying to keep the rest of the inputs constant in the different shots. To be able to analyse the viewer's neuronal response to a specific cinematographic technique in the most concrete possible way, Heimann introduced differences only in one variable in the filmed audiovisual material that was shown to the viewer.

However, in the present study, we set out to work with existing feature films, so we face the challenge to isolate the input of the cut of all the non- related features that appear in the film after the shot change by cut. In this way, we consider that we can perform an analysis of the film edition closer to the real cinematographic experience than working from material filmed *ad hoc* for laboratory purposes.

3 EEG Registration

The experimental designed consisted on the selection of four film fragments belonging to commercial films together with the design of the electroencephalographic record. The film fragments contain the shot changes and represent different cinematographic styles in order to have the widest possible representation of reactions due to different technical, aesthetic and stylistic characteristics of the films. Subjects who participated in the experiment visualized the film fragments while we recorded their electroencephalogram. The selected film fragments belong to *Bonnie & Clyde*, *The Searchers*, *Whiplash* and *On the Waterfront*.

Twenty-one subjects participated as spectators. The participants did not report any neurological disorder, psychological problems or being under medication. The subjects were kindly asked not to consume exciting substances (such as coffee) or depressants (such as alcoholic drinks) that could modify cognition in the hours before the experiment.

The EEG signal from the 21 volunteers was recorded using a 31 electrodes array distributed over the scalp following positions defined by the international convention of the American Electroencephalographic Society in its 10–20 system for EEG-MNC [10] (Fig. 1).

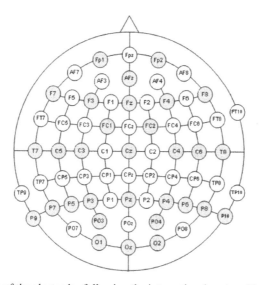

Fig. 1. Distribution of the electrodes following the international system 10–20 for EEG-MNC.

Electroencephalographic registers were made between 15 and 30 of May 2017. The four film fragments were played to each subject through the Unity game engine [11], while the electroencephalogram was registered in MatLab though two different computers synchronized by means of a User Datagram Protocol Network (UDP).

4 Signal Analysis Design

In order to identify neuronal patterns triggered by the event of the cut in the EEG signal we divide the analysis in two phases. The first phase consists in preparing the signal registered by the EEG for the study in the frequency domain. The second phase consists in the statistical analysis in the power change of the signals obtained to locate reaction patterns produced as a result of the shot change by cut.

4.1 Signal Preparation

After recording the electroencephalogram an ICA (independent component analysis) was performed in order to isolate for each electrode the signal from its registration area, discarding the interferences from other electrodes recording areas. Then, we proceeded to perform a manual cleaning of artefacts from the recorded signals. From there, we generated one model of neuronal reaction signal for each of the audiovisual fragments yielding four models. To do this, we divided the continuous signal recorded for each user into the four sections of interest corresponding to the recording of each audiovisual clip shown, one for each film, eliminating the rest of the signal. Using the MatLab G.bsanalyze toolbox, we averaged the signals for each electrode in each audiovisual fragment for all 21 subjects [12]. Therefore, we ended up with a set of four model signals, one for each of the audiovisual clips and for each of the 31 electrodes averaged across the 21 spectators analysed. The scheme of the process of creating the model signals is shown in the Fig. 2.

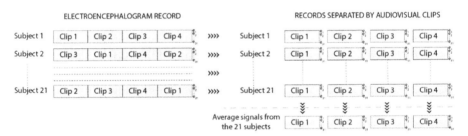

Fig. 2. Scheme of signal registration.

Given that the regions of interest for our study are the instants after the cut, from each model we selected only the 2000 ms time intervals centred in each cut (one second after and one before) and dismiss the rest (left part of the Fig. 2). In this way, we ended up with 84 two-second epochs for *Bonnie and Clyde*, 15 for *The Searchers*, 112 for *Whiplash* and 47 for *On the Waterfront*. These signal fragments for each cut are combined with G.bsanalyze for each audiovisual fragment, obtaining a model signal of the neuronal reaction for the cut for each audiovisual fragment in each electrode. From this process we obtain 31 model signals for each film fragment, one for each electrode, which we will call ASF (Average Signal by Film). The generation of ASF model signals from all users and cuts for each film fragment keeps the common reactions, minimizing the specific ones of the users and minimizing also the different kinds of cuts or the aspects not related to the common event of the shot change. Through this signal processing we have for each electrode four different model signals of reaction to the cut, one for each film fragment, thus being able to compare them and discard what is not common to the cut, such as narrative, aesthetic or technical characteristics of each film fragment. In Fig. 3 we can see graphically the scheme of the creation of ASF signals from the model signals obtained for each audiovisual clip from the electroencephalographic records obtained from the 21 users.

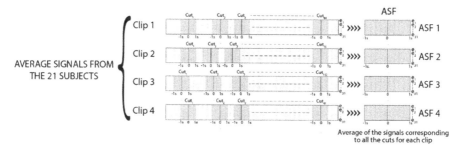

Fig. 3. Scheme of obtaining ASF signals.

The ASF model signals obtained are transformed to the frequency domain. For this transform we take sequences of 16 successive samples for each signal in the EEG register timeline (taken at 256 Hz sampling) with the toolbox G.bsanalyze. In this way we obtain the power change for each time interval of 62.5 ms in the frequency domain from the fast Fourier transform. This value called power change represents the oscillation frequency in the EEG register, allowing us to analyse the modulation of neuronal rhythms, corresponding to the electrode and frequency range analysed in response to the experiment stimulus [13, 14]. In our specific case we can know the excitation and inhibition of neurons in different neuronal areas in response to the shot change by cut. Power change values are collected in 4 matrices of dimension 31×32, one matrix for each ASF. In each matrix the rows correspond to the electrodes and the columns to the temporal progression representing 62.5 ms intervals.

The power change analysis allows to determine the event-related desynchronization/synchronization. To analyse the responses of each signal in the power change we define the frequency bands in: 0.5–3 Hz for Delta, 3–7 Hz for Theta, 7–14 Hz for Alpha, 14–32 Hz for Beta and 32–42 Hz for Gamma. In addition to this general division, we also perform another segmentation differentiating High and Low ranges of each frequency band. Thus we establish the ranges 0.5–1.5 Hz for Low Delta, 1.5–3 Hz for High Delta, 3–5 Hz for Low Theta, 5–7 Hz for High Theta, 7–10.5 Hz for Low Alpha, 10.5–14 Hz for High Alpha 14–23 Hz for Low Beta, 23–32 Hz for High Beta, 32–37 Hz for Low Gamma and 37–42 Hz for High Gamma. We can see a scheme in the Fig. 4.

From this process we obtain a model signal of reaction to the cut for each electrode in each of the 15 frequency bands indicated from each ASF corresponding to a film fragment. We get 15 matrices for each of the 4 film fragments, corresponding to the power change values in a specific frequency band, resulting in a total of 60 matrices. Each matrix has 31 rows and 32 columns. Each row being the Power Change values for

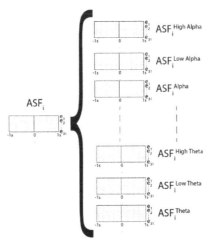

Fig. 4. ASF breakdown by frequency ranges.

each electrode in a given frequency band along the selected two seconds of time. The first 16 columns correspond to the values of the second before the cut and the next 16 columns corresponds to the signal after the shot change.

These power change matrices are represented in Eq. 1, where M_i^j represents the matrix for the frequency band j and the ASF i. The variable j can take values from 1 to 15, one for each frequency range, and the variable i take values from 1 to 4, one for each audiovisual fragment. Matrix values are represented in Eq. 1 as Electrode$x$ Time$y$ representing the power change value for each electrode along the time. The subscript of each electrode (Ex) is between 1 and 31, while the subscript of time (Ty) is between 1 and 32.

$$M_{ASF\,(i[1,4])}^{Banda\,de\,frecuencia(j[1,15])} \begin{pmatrix} Eectrodo_1\,Tiempo_1 & \cdots & E_1 T_{32} \\ \vdots & \ddots & \vdots \\ E_{31} T_1 & \cdots & E_{32} T_{32} \end{pmatrix} \tag{1}$$

4.2   Treatment of the Data Obtained in the ASFs

The next phase of the methodology consists on comparing the data obtained in the ASF. The comparative processes applied on the Power Change of the ASF in the different frequency bands serve to locate those neuronal reactions that have been triggered due to the cut event. In the following two sections we first explain the design of the ASF signal analysis, and then describe its application. Here we define a methodological procedure to locate neuronal reactions due to the event of the shot change by cut.

ASF Signal Analysis Design. As we designed the experiment with real film fragments, extracted from feature films destined to the general public, in addition to the shot changes by cut there are a great diversity of uncontrolled stimuli, which complicates the identification and isolation of the neuronal reactions due to the cut. To determine which responses are specifically associated with the shot change cut, which is the common event in all the studied signals, we analyse the correlations between the power change for each possible combination of a specific electrode between the four ASF, during the same period of time and in the same frequency band. Thus, to locate in the second after the cut only those reactions associated with the common event of the cut, we search those time intervals where correlations happened between all possible combinations by pairs between the four ASF signals for the same electrode in the same frequency band. This methodology follows the idea proposed by Lachaux, Chavez and Lutz [12], with the difference that their investigation seeks to locate the correlation between different frequency bands in the same stimulus, while we are interested in seeking correlation in the same frequency band, isolating the reaction to the studied stimulus.

It has become customary in neurocinematic studies to compare between the signals recorded by the same electrode in the same frequency band for similar events in order to analyse the neuronal response in the ERD/ERS [2, 5], mainly using ANOVA (analysis of variance). The difference in our approach is that we want to locate the similarities between signals, so we resort to comparison methodologies that detect dependency structures and correlations between the samples instead of resorting to analysis of variance techniques. Another variation that we add in our approach with respect to the referred works is that we want to obtain results in temporal continuity, so instead of establishing watertight periods of time we preformed all analyses with sliding temporal windows.

This dependence and correlation between different ASF signals is the necessary condition to assume that an electrode in a specific frequency band represents a neuronal response due to the cut event, but at the same time, it's not sufficient to determine the neuronal activity as significant. Then, it is necessary that in addition to the dependence and correlation between the signals of the different ASF, each one contains a significant variation in the power change evolution, indicating that there is a synchronization or desynchronization in the neuronal rhythms (ERD/ERS). This second process discards those dependencies and correlations between signals that are not associated with a significant variation of the power change representing little variation in the neuronal rhythms of the EEG register. The result finally obtained is the location of the temporal windows, the electrodes and the frequency bands that allow us to locate the significant neuronal responses triggered by the event of the shot change by cut.

Procedures for the Search of Neuronal Reactions Due to the Cut Event. Our methodological approach is based on comparing the power change samples between the four ASF through sliding time intervals for 31 electrodes in each one of the 15 frequency bands. In order to apply it, we establish a work routine using MatLab, which allows us to develop an automated analysis system. To develop the comparison processes between peers of signals, we assign a binary value based on the degree of dependence and correlation for each comparison between the different ASFs for the same electrode in a certain frequency band in a time interval. In this way we assign the value of 1 when a comparison overcome a certain representative threshold value. Comparisons between peers of signals imply a test to analyse the adjustment of depending structures using the permutation test (p-value) and to analyse the correlation between samples through the Spearman test (Rho). These peer comparison tests are applied at time intervals as slide temporal windows. This approach allows us to obtain results with temporal continuity that collect the neuronal reactions to the shot change in the different electrodes and frequency bands.

The comparisons test is made between sliding windows with a range of 6 consecutive power change values. This range is equivalent to 375 ms, because each sample represent an interval of 31.25 ms. The choice of 6 samples to conform the temporal windows responds to the use of the permutations test, due to the fact that if the intervals have less than 5 samples the results can lose robustness [13], and if we have more sampling we would obtain less temporal resolution. Consequently, we segmented the 2 s around the cut from the registration of each ASF (2000 ms) in 27 temporal windows of 6 correlative samples of each temporal window. In this distribution the window consisting of the three temporal samples immediately before and after the cut have the time value 0 and represent the cut.

Equation 2 shows the transformation of the matrix containing the power change values into matrices that contain sliding window vectors. Each element of the resulting matrix, instead of assuming a value as we had before, contain a vector of 6 values. Each vector consists of 6 samples from the power change corresponding to each sliding window. We call it vEkVTz to indicate that it is a vector (v), where the row position indicates the electrode (Ek) and the column position the corresponding time window (VTz). The resulting matrix continues having 31 rows, one for each electrode, but have 27 columns, one for each time window. In consequence, the 14th temporal window (vEkVT14) corresponds to the instant of the shot change by cut, because it is formed by the three samples before the cut (EkT13, EkT14, EkT15) and the three after (EkT16, EkT17, EkT18). Resulting for example vE1VT1 = (E1T1, E1T2, E1T3, E1T4, E1T5, E1T6) and vE31VT27 = (E31T27, E31T28, E31T29, E31T30, E31T31, E31T32).

$$
M_{i[1,4]}^{j[1,15]} \begin{pmatrix} E_1 T_1 & \cdots & E_1 T_{32} \\ \vdots & \ddots & \vdots \\ E_{31} T_1 & \cdots & E_{31} T_{32} \end{pmatrix} \rightarrow M_{i[1,4]}^{j[1,15]} \begin{pmatrix} vE_1 VT_1 & \cdots & vE_1 VT_{27} \\ \vdots & \ddots & \vdots \\ vE_{31} VT_1 & \cdots & vE_{31} VT_{27} \end{pmatrix} \tag{2}
$$

The identification of dependencies and correlations between signal pairs consists of two phases. The first phase is to apply the permutation to locate these neuronal reactions that show a dependency structure between signals. The second phase consists in using Spearman's test to know the level and type of correlation between the two signals. By applying both processes we seek to be restrictive in the detection of temporal windows that we can consider as neuronal reactions to the shot change by cut, since the whole process proposed is aimed at isolating the neuronal responses of the cut event from a random set of stimuli derived from the reception of the film fragments showed.

To put it into practice, the analysis applied to each pair of time windows is performed in the two phases described. In the first phase we take two ASFs in a specific frequency band and compare for the same electrode the same time window formed by 6 power change samples. We compare this pair of signals through the permutation test. Setting the threshold of significance at 0.05 [17] we indicate with the value 1 when the result shows a significant dependent structure ($p < 0.05$). Equation 3 shows how this process operates. The permutations test (PT) is applied to the electrode k for the power change samples included in the time window z (VTz) in the frequency band j ($\forall j \in [1, 15]$).

This comparison is made between two different ASFs, so that i and i' have different values between 1 and 4 ($\forall$ i, i' $\in$ [1, 4]: i $\neq$ i'). With 4 ASFs there are 6 possible pairwise comparisons between i and i'.

$$PT_{i,i'}^{j}\left[M_i^{j}(vE_kVT_z)\,;M_{i'}^{j}(vE_kVT_z)\right] \begin{cases} \textit{If } p < 0.05 \;\rightarrow\; PT_{i,i'}^{j}(E_kVT_z) = 1 \\[2mm] \textit{If } p \geq 0.05 \;\rightarrow\; PT_{i,i'}^{j}(E_kVT_z) = 0 \end{cases} \tag{3}$$

If the result obtained shows a significant p-value ($p < 0.05$) it is considered acceptable and is indicated as 1 in a corresponding matrix position where the results are indicated. This technique is an application of the image segmentation methodology used in tomography [18], but we applied it to the segmentation of the range of values resulting from the application of the peer permutation test between 0 or 1, setting the p-value as threshold. The resulting matrix contains all the comparisons made by the permutation test for the same frequency band between two ASFs. Each row corresponds to an electrode and each column to a temporal window.

In this way we detect the temporal windows where the compared data have a significance higher than the p-value threshold, revealing a dependency structure between the compared samples. As a result, we obtain 6 matrices for each frequency band, each one containing the comparisons between two of the 4 ASFs. In the case that in these 6 matrices for a same frequency band, in a time interval of the same electrode it shows a significant dependence, it is considered that this time interval in that electrode and frequency band is reflecting a reaction that could be identified as a neural reaction due to the stimulus of the shot change by cut. In this way, we obtain as a result a final matrix for each frequency band where it is indicated in which electrodes during the same time window occurs a significant dependence relationship in all possible comparisons between the 4 ASFs. The process description is shown in Eq. 4. When in a temporal window and electrode (EkVTz) in the same frequency band (j) occurs that the sum of the pairwise comparisons Boolean results of the 4 ASF matrices i e i' ($\forall$ i, i' $\in$ [1, 4]: i $\neq$ i') results in the value of 6, it means that all the ASFs shows a dependency relationship in that time window and electrode for the defined frequency band. Therefor we assign the value 1 in the corresponding position of a new matrix called PTTotalj, representing j the specific frequency range.

$$PTTotal_{31 \times 27}^{j} = \sum_{i,i' \in [1,4]:i \neq i'} PT_{i,i'}^{j} \begin{cases} \textit{If } \sum\limits_{i,i' \in [1,4]:i \neq i'} PT_{i,i'}^{j}(E_kVT_z) = 6 \;\rightarrow\; PTTotal^{j}(E_kVT_z) = 1 \\[4mm] \textit{If } \sum\limits_{i,i' \in [1,4]:i \neq i'} PT_{i,i'}^{j}(E_kVT_z) < 6 \;\rightarrow\; PTTotal^{j}(E_kVT_z) = 0 \end{cases} \tag{4}$$

Applying this procedure to all temporal windows we obtain those that for each electrode and frequency band show dependence between the different film fragments, where the only common event between the 4 ASFs is the shot change by cut. The resulting matrix from this process have a dimension of 31 × 27. The rows correspond to the electrodes and the columns correspond to the temporal windows. We obtain a total of 15 Boolean matrices, one for each frequency band (j).

Once this stage of the process is finished, in the second phase we proceed to determine the goodness of the correlations in the temporal windows detected as dependent. To do this, we apply the Spearman correlation test in the temporal windows where a dependency relationship has been detected by the permutation test. The objective is to verify that the peer relationships established between the same temporal windows of the same electrode in a given frequency band, in addition to showing a dependency relationship, are also correlated. To process it in MatLab, the double condition that the permutations test and the Spearman test is met is easier to obtain separately PTTotal j (Eq. 4) and an equivalent Boolean matrix by applying the Spearman correlation test and compared them.

The Spearman test is applied similarly to the permutations test, establishing pairwise comparisons for each time window of each electrode between the different ASFs in the same frequency range. If the correlation value Rho exceeds the threshold value, it is indicated as 1 in the resulting matrix. To define the correlation between the samples we set the Rho threshold value at 0.5 as stated in abundant literature [19, 20]. The type of correlation that we are looking for among ASFs is positive, since what we are looking for is to detect similar and non-opposite responses. We discard all the Rho negative values because it represents inverse correlations, remaining exclusively the Rho values greater than 0.5 and not those under −0.5, using the Spearman correlation test in one tail form [21]. Equation 5 represents the explained process. For the compared pairs of temporal windows (VTz) of two ASF (between i and i′ being i different from i′) on the same electrode (Ek) in a certain frequency band (j). In case of obtaining a Rho value greater than 0.5 means a positive and significant correlation, then we indicate the result as 1 in the matrix $SMR^j_{i,i'}$ in the corresponding matrix position (EkVTz).

$$SMR^j_{i,i'}\left[M^j_i(vE_kVT_z)\,;\,M^j_{i'}(vE_kVT_z)\right]\begin{array}{l} \text{If Rho} > 0.5 \rightarrow SMR^j_{i,i'}(E_kVT_z) = 1 \\[2em] \text{If Rho} \leq 0.5 \rightarrow SMR^j_{i,i'}(E_kVT_z) = 0 \end{array} \qquad (5)$$

As we did with the permutation test, we identified as positive only the correlated reactions that exceed the Rho threshold in the 6 possible comparisons between the 4 ASFs, discarding weak correlations. In Eq. 6 we describe this process. When the sum of the results obtained in the same temporal windows for the same electrode (EkVTz) between the different Spearman correlation matrices for each pair of comparison ($\forall$ i, i′ $\in$ [1, 4]: i $\neq$ i′) of the same frequency band (j) results in a value of 6, means that among all the ASF is a correlation at that moment, in this electrode for that frequency band, so we assign the value 1 in the corresponding position of the resulting matrix SMRTotal j.

$$If \sum_{i,i' \in [1,4]: i \neq i'} SMR_{i,i'}^{j}(E_k VT_z) = 6 \rightarrow SMRTotal^j(E_k VT_z) = 1$$

$$SMRTotal_{31x27}^{j}(E_k VT_z) = \sum_{i,i' \in [1,4]: i \neq i'} SMR_{i,i'}^{j}(E_k VT_z) \qquad (6)$$

$$If \sum_{i,i' \in [1,4]: i \neq i'} SMR_{i,i'}^{j}(E_k VT_z) < 6 \rightarrow SMRTotal^j(E_k VT_z) = 0$$

The matrix SMRTotal j it allows to identify the temporal windows in which an electrode in a specific frequency band shows correlation in the power change between all the film fragments used. From the whole process we obtain a total of 15 matrices of dimension 31×27, where each matrix corresponds to a specific frequency band. In these matrices the rows correspond to the 31 electrodes and the columns to the 27 temporal windows. Thus, we take into consideration only the temporal windows for each electrode in each frequency band where there is a representative dependence relationship and a correlation representative in the recorded signal, reflecting the common event of the cut. We accept these results as evidence of neuronal reactions in the power change due to the shot change by cut, which is the only common event in the 4 film fragments showed.

Next, in order to compare the permutations test and the Spearman correlation test, as both matrix are Boolean, we proceed to multiply the values that occupy the same matrix positions between the matrices PTTotal j y SMRTotal j in the same frequency band (j). In this way, we obtain a Boolean result matrix that meets the double condition of dependence and correlation, indicated as 1 only the temporal windows where p-value is less than 0.05 and Rho greater than 0.5 in all possible combinations between entre i and i' ($\forall i \neq i'$). In Eq. 7 we indicate the multiplication operation between the values that occupy the same matrix position as ".x", similar to the execution command in MatLab (.*), to differentiate the notation used from a standard matrix multiplication. The result of this operation is the location of the temporal windows where there is correlation and dependence between the 4 ASFs for the same electrode in the same frequency band (j). That is, the value 1 identifies in the resulting matrix all those temporal windows where a neuronal reaction due to the event of the cut is detected. We call this final matrix CD (Cut Dependencies).

$$CD_{31x27}^{j} = PTTotal_{31x27}^{j} .x\ SMRTotal_{31x27}^{j} \qquad (7)$$

The resulting matrix CD j has a dimension 31×27, where each row represents an electrode and each column a temporal window. 15 matrices are obtained, each for an analysed frequency band (j). In these matrices, the temporal windows of each electrode that show a neuronal response due to the cut event contain the value 1, while the rest contain the value 0.

The data obtained from the proposed analysis allows a temporal study of the neuronal response triggered by the cut event for each frequency band and electrode. Therefore, the last step in the methodology is to identify those temporal windows that, in addition to showing reactions triggered by the change of plane by section, contain significant variations in the temporal evolution of the power change. For this we resort to the analysis of the slopes of the signals [22, 23]. This process is performed by calculating the slopes separately in the different time windows, after their transformation into a Boolean matrix and finally multiplying the values that occupy the same position of the matrix obtained with the also Boolean matrix CD, already calculated above. In this way, as we operate with Boolean matrices, we get 1 when the temporal windows detected as dependent on the cut event contain a slope that exceeds the threshold and maintains the same sign in the 4 ASF.

Variations in the evolution of power change indicate a process of synchronization or desynchronization in neuronal rhythms [24, 25]. Therefore, if we detect significant slopes in the temporal windows from an electrode at a frequency band in which we have detected a possible neuronal response due to the shot change by cut, we obtain as a result the location of important synchronization or desynchronization processes as consequence of the cut. Then, for all temporal windows of each electrode in each frequency band where dependence and correlation has been detected in the matrix CD j(EkVTz) we analyse the slopes in the power change values of the 4 ASFs.

To perform this process, we take each vector formed by the 6 temporal samples corresponding to the time window (vEkVTz) in each of the 4 ASFs and we calculate their slope. If the slopes calculated in the temporal windows exceed a representative threshold in the 4 ASFs with the same sign, the result is marked as 1, indicating all others as 0. It is important to consider the sign of the slope, because we can have positive (ERS) or negative (ERD) slopes, and only results that show the same type of variation in neuronal rhythms for the 4 ASFs in the same time window of the same electrode and in the same frequency band can be representative. Therefore, we consider as relevant in terms of variation of the power change only those slopes that exceed a certain threshold with the same sign in the same time window, electrode and frequency band for the 4 ASFs.

To establish a threshold value that defines the level of variation to due to the shot change by cut as representative, we evaluate the slope of the power change. This evaluation must be weighted in relation to a state of rest or balance. To establish the threshold value of the slope that identifies such variations of the power change as representative, we analyse the data contained in the second before the shot change (baseline) as a reference. When performing operations with sliding windows of 6 samples from the power change, we obtain 27 temporal windows in the 2 s analysed, one second previous to the cut and one second after. From these temporal windows we consider the intermediate one as the instant of the cut, because it is composed by 3 samples before and three after the shot change. This means that the two temporal windows prior to the intermediate one will also contain samples after the cut (Fig. 5), so if we consider them as baseline, they can contaminate the reference previous to the stimulus with neuronal reactions after the shot change. Therefore, in order to establish the threshold value that determines the representative slopes in the temporal windows of the power change, we use the first 11 temporal samples as a baseline, excluding the temporal windows 12 and 13, as we can see in the Fig. 5.

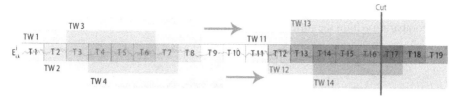

Fig. 5. Temporal sliding windows that are not influenced by power change samples after the shot change by cut.

In neurological studies, the segmentation of data in "below average" and "above average" is usually used in order to discriminate between two groups depending on a specific characteristic that is considered to be more represented in one of the two parts than what would be the case in the whole data set [26–28]. In our case, we take as significant slopes in the temporal windows of the baseline those above average. Therefore, we set the threshold value as the average obtained in the defined baseline. To calculate the average, we analyse the slopes in the sliding windows of the baseline formed by the first 11 sliding windows that do not contain samples after the shot change by cut. From the baseline sliding windows, we select as representative those that show a common behaviour, eliminating all those windows that for the same electrode, in the same frequency band and in the same time window do not have the same sign in the 4 ASF. We only consider those cases in which the slope has the same increasing or decreasing trend in the 4 ASF. We carry out this previous operation to obtain the average data in a similar way to how the detection of the significant slopes on the temporal windows will be worked after the cut. We are interested in knowing the average inclination in degrees to calculate the deviation either positive or negative with respect to $0°$, so we need to know the average inclination at its absolute value. For this we calculate the average slope from the absolute values of each slope and thus finally obtain the threshold value that determines whether a slope value is significant.

Once the threshold of significance for the slopes is established, we proceed to operate in the different temporal windows in order to know the inclination of the neuronal reactions due to the event of the cut. To do this we begin by calculating the slopes of each time window, as defined in Eq. 8. As we calculate the slope in 6 sample windows, the denominator is 6 and the samples at the ends of each time window are Tz y $Tz + 5$ for each frequency band (j) in each ASF (i).

$$Slope_i^j(vE_k VT_z) = actan\left(\frac{E_k T_{z+5} - E_k T_z}{6}\right) \tag{8}$$

We create a matrix $Sl_i^j(E_k VT_z)$, showed in Eq. 9, from the matrix $Slope_i^j(vE_k VT_z)$, which contains the values of the slopes for each specific frequency band (j) in a specific ASF (i) for each electrode (Ek) and each time window (VTz), to calculate the slope from sliding windows of 6 temporal samples on each electrode (vEkVTz). Once the slopes are calculated, if they exceed the positive threshold they are indicated in the matrix Sl as 1, and if they exceed the negative threshold they are indicated as -1. The rest of the matrix positions are indicated as 0. This differentiation between 1 and -1 will allow us to ensure that the slopes of each time window have the same slope sign in the 4 cases when comparing within the different ASFs.

$$Sl_i^j(E_kVT_z) \quad
\begin{cases}
\text{If Slope } vE_kVT_z > \text{Threshold } (+) \rightarrow Sl_i^j(E_kVT_z) = 1 \\[2mm]
\text{If Threshold } (-) \leq \text{Slope } vE_kVT_z \geq \text{Threshold } (+) \rightarrow Sl_i^j(E_kVT_z) = 0 \\[2mm]
\text{If Slope } vE_kVT_z < \text{Threshold } (-) \rightarrow Sl_i^j(E_kVT_z) = -1
\end{cases}$$

(9)

In this way we obtain, as indicated in Eq. 9, the identification of all the slopes that exceed the threshold in a time window. This process results in 15 matrices for each film fragment (i), each one corresponding to its frequency band (j). The matrices Sl have a dimension 31×27, the rows corresponding to the electrodes and the columns corresponding to the temporal windows.

When applying the slope analysis, it is important to ensure that we locate those whose variation in the power change represent a monotonous behaviour, that is, whose slope calculated from the first and last value of the time window of 6 temporal samples is consistent with the trend of the slope throughout the interval. For this, we verify that within each window of 6 temporal samples where a representative slope is located, that window also contains in its samples at least an interval with a slope of 3 samples that is also representative and have the same sign. If we detect that the slopes are not monotonous we change the value 1 for 0 in the corresponding matrix position.

Once the power change slopes are known for each time window, we proceed to compare each one with their equivalent in the same electrode and frequency band for the different ASFs. In Eq. 10 we can see how the matrix SlopeTotal j referring to the significance of the slopes in the power change between the different ASFs is obtained for each frequency band j, which later we multiply, position by position, by the corresponding matrix CD j previously obtained (Eq. 7). This procedure allows us to locate the temporal windows that show a correlation and dependence between the neuronal reactions triggered by the shot change by cut, which have an important variation in the power change, reflecting a variation in the neuronal rhythms that we will analyse through the ERD/ERS.

The matrix SlopeTotal j is obtained through the sum of each Boolean value assigned to the calculation of the slope made on the power change in the 4 ASFs (i) in windows composed of 6 temporal samples (VTz) for the same electrode (Ek) in the same frequency band (j). As previously explained, each slope of a temporal window that exceeds the defined threshold is indicated as $+1$ or -1. To transform the matrix SlopeTotal to Boolean and be able to operate it with the matrix CD, we take only as acceptable the temporal windows in which all the slopes for the same frequency band in the same electrode for the 4 ASFs have an inclination that exceeds the threshold, so that the optimal results of the summation should be 4 or -4. Each matrix position where the result of the sum is 4 or -4 is replaced by 1. All the other positions are indicated as 0 because they are not relevant. The process is explained in the Eq. 10.

$$If \sum_{i=1}^{4} Sl_i^j(E_k VT_z) = 4 \rightarrow SlopeTotal^j(E_k VT_z) = 1$$

$$SlopeTotal_{31x27}^j = \sum_{i=1}^{4} Sl_i^j \longrightarrow If \sum_{i=1}^{4} Sl_i^j(E_k VT_z) \neq -4 \; (\&) \neq 4 \rightarrow SlopeTotal^j(E_k VT_z) = 0$$

$$If \sum_{i=1}^{4} Sl_i^j(E_k VT_z) = -4 \rightarrow SlopeTotal^j(E_k VT_z) = 1$$

$$(10)$$

In Eq. 10, we obtain a Boolean matrix (SlopeTotal j) of dimension 31×27 that identifies whether there is an instance in a specific time period of the power change that exceeds a certain threshold with the same sign on the same electrode for the 4 ASFs in the same frequency band (j). The rows of the matrix SlopeTotal correspond to the 31 electrodes and the columns to the 27 temporal windows.

Once the matrix SlopeTotal j is obtained for each frequency band j, we proceed to combine it with the matrix CD j previously obtained. In this way we locate those temporal windows that have a synchronization or desynchronization process due to the event of the cut. The process is explained in Eq. 11. To combine these matrices, we multiply the values that occupy the same matrix positions (.x) between SlopeTotal j and CD j for the same frequency range (j). Through this process we obtain the matrix DSC j (Desynchronizations and Synchronizations due to the Cut) for each frequency band (j).

$$DSC_{31x27}^j = CD_{31x27}^j .x \; SlopeTotal_{31x27}^j \tag{11}$$

The matrix DSC obtained in Eq. 11 is a Boolean matrix where each time window is indicated as 1 if a neuronal reaction triggered by the cut event has been detected and shows a significant process of synchronization or desynchronization. We obtain 15 matrices DSC, one for each frequency band. Each matrix DSC has a dimension 31×27, where the rows correspond to the electrodes and the columns to the temporal windows.

When combining the results of the monotonous slopes with the results obtained by permutation test and correlations of Spearman, all made with sliding windows of 6 samples, we are sure that the representative slopes that occur in the 4 ASFs for the same time in the same electrode and frequency band are related. Therefore, we performed a security check as similar as possible, whether for the analysis of the slopes in the windows of the three temporal samples or for the six temporal samples. We cannot reliably apply the permutation test on three-sample windows [16], but we can apply the Spearman correlation test and get acceptable results [29]. In this way, in addition to verifying that the behaviour of the slope is monotonous, correlated and dependent on the slope for the temporal windows of 6 samples, we also verify that the relationship between the 4 ASFs is maintained despite reducing the size of the sliding window. That is, when we detect dependence, correlation and a significant slope in a period of 6 samples among the 4 ASFs, we ensure that a segment of 3 samples contained in this period also shows correlation and a significant slope.

In summary, we consider that a specific temporal window of an electrode in a given frequency band reacts significantly due to the shot change by cut, either in the form of synchronization or desynchronization, when three established conditions are met: overcome the p-value in the permutation test, overcome the Rho in the Spearman's correlation and have a monotonous slope that overcome the defined threshold. All the mathematical processing to get the results is carried out automatically by means of scripts executed by MatLab. Thanks to programming these scripts we can perform a depth analysis that automatically locates the significant reactions in the power change due to the cut event.

4.3 ERD/ERS Study of Temporal Windows Identified as Dependents to the Cut

Once the neuronal reaction triggered by the cut event are located knowing each temporal window for electrode and frequency band, we proceed to analyse its ERD/ERS from the registered power change. The analysis of the ERD/ERS is based on the detection of variations in neuronal activity in relation to the state of non-reaction to the event (baseline) to be able to make an observation in terms relative to a state prior to the variation of the neuronal rhythms. This variation is calculated in percentage terms with respect to a state of no reaction to the analysed event.

To quantify the variation of the power change that occurs in an electrode in a certain frequency band we do it with respect to the average value of the signal in the second prior to the shot change by cut (baseline) [9]. To obtain the value of the ERD/ERS we apply Eq. 12 [24, 25]:

$$ERD = \frac{Baseline - Test}{Baseline} \times 100 \qquad (12)$$

Applying Eq. 12 we obtain the ERD/ERS values as a percentage. Resulting from this operation, positive values correspond to desynchronization processes (ERD) and negative values to synchronization process (ERS). This system allows us to understand the evolution of the neuronal rhythms in the spectator facing the shot change by cut.

5 Discussion

The application of the methodological design though automated scripts in MatLab allowed us an intensive and complete analysis for all the electrodes in all the frequency ranges over slide time windows, allowing us to locate and identify any variation that affects Power Change as a consequence of the event of the cut. This allows us to be able to perform an in-depth analysis for all the data registered through the EEG, without having to limit the analysis to specific data biases and also avoids the need to produce *ad hoc* audiovisual material for the laboratory [2]. Until now, for the analysis of the spectators by means of the EEG, audiovisual clips have been produced specifically for laboratory conditions in order to assess specific cinematographic technical aspects [2, 8, 9]. Fragments extracted from real films have been used only to study emotions where specific technical aspects of the film were not taken into account [30, 31]. The methodology developed in the present investigation allows us to obtain successful results concentrating on cinematographic techniques from films fragment originally designed for the consumption of cinematographic spectators, allowing to make a contribution to the cinematographic theory based on the analysis of the film itself and not from audiovisual recreations filmed for laboratory purposes.

In the methodology two processes are established. A first one oriented to detect all the temporal windows for each electrode and frequency band where neuronal reactions are a consequence to the shot change by cut, and a second process where we identify which of these neuronal reactions constitute representative variations in the neuronal rhythms. Following the exposed methodology we can locate the event of the shot change by cut recognizing modulation patterns in the frequency bands Theta and Delta. We can identify activity in different cortical areas the first 250 ms after the cut, most of which are Theta synchronization. In addition, we can locate a clear theta synchronization tendency between 700 ms and 1000 ms, which coincides with a desynchronization of Delta in the same period. Most of the activity found is localized in the parietal area. The results obtained indicate a clear involvement of the hippocampus in the event of the shot change by cut, which supports our methodology because it coincides with the results of previous investigations through fMRI [1].

6 Conclusion

The present study proposes a novel methodological approach for the detection of neuronal responses triggered by the shot change by cut. The use of real film fragments that are stylistically diverse for the study of the EEG on aspects related to the cinematographic technique requires debugging all those reactions due to the film context, allowing to be able to analyse films designed for the spectator and not audiovisual clips filmed for the experiment itself. The exposed methodology for the detection of temporal windows in electrodes and frequency bands in which neuronal reactions registered in the power change occur due to the stimulus of the cut, although it is very restrictive, it allows us to extract clear results specifically associated with the perception of the shot change by cut which, at the same time, are backed by previous investigations [1]. The restrictiveness of the designed system means that we only consider 1.02% of temporal windows, which are those dependent and correlated between the 4 ASFs, and 0.02% which are dependent and correlated with significant variations in their Power Change between the 4 ASF.

To adapt Brain-computer interfaces to real world situations we need work with noise situations. Stimuli from audiovisual clips are complex as real life ones. Previous experiments works under the noise absence's premise. Therefore, methodologies involved in events detection thru EEG analysis need to transcend simplicity of those experiments. Our approach shows how it is possible to isolate a particular kind of event using the proposed framework. To find brain response, which a quick EEG analysis can process almost in real time, it is necessary to previously identify the patterns associated with the events. We show here a working framework adaptable to many events detection that need to identify time lapses and brain regions involved in human response to such events.

References

1. Ben-Yakov, A., Henson, R.: The hippocampal film-editor: sensitivity and specificity to event boundaries in continuous experience. bioRxiv, 273409 (2018)

2. Heimann, K., Uithol, S., Calbi, M., Umiltà, M., Guerra, M., Gallese, V.: "Cuts in action": a high-density EEG study investigating the neural correlates of different editing techniques in film. Cogn. Sci. **41**(6), 1–34 (2016). https://doi.org/10.1111/cogs.12439

3. Smith, T.: An attentional theory of cinematic continuity. Projections **6**, 1–50 (2012)

4. da Silva, B., Bandeira, A., Tavares, M.: Cadavre exquis: a motion-controlled interactive film. In: 9th International Conference on Digital and Interactive Arts, Braga, Portugal, pp. 1–4 (2019)

5. António, R., da Silva, B., Rodrigues, J., Tavares, M.: Experimenting on film: technology meets arts. J. Creat. Interfaces Comput. Graph. **8**(1), 54–56 (2017). https://doi.org/10.4018/IJCICG.2017010104

6. Cohendet, R., da Silva, M., Gilet, A.-L., Le Callet, P.: Emotional interactive movie: adjusting the scenario according to the emotional response of the viewer. EAI Endorsed Trans. Creat. Technol. **4**(10), 1–7 (2017). https://doi.org/10.4108/eai.4-9-2017.153053

7. Nijholt, A.: Brain Art: Brain-Computer Interfaces for Artistic Expresion. Springer, Cham (2019). https://doi.org/10.1007/978-3-030-14323-7

8. Heimann, K., Umiltà, M., Guerra, M., Gallese, V.: Moving mirrors: a high-density EEG study investigating the effect of camera movements on motor cortex activation during action observation. J. Cogn. Neurosci. **26**(9), 2087–2101 (2014). https://doi.org/10.1162/jocn_a_00602

9. Martín-Pascual, M.A.: Mirando la realidad observando las pantallas. Activación diferencial en la percepción visual del movimiento real y aparente audiovisual con diferente montaje cinematográfico. Un estudio con profesionales y no profesionales del audiovisual. (tesis doctoral). Universitat Autònoma de Barcelona, Barcelona, Spain (2016)

10. American Electroencephalographic Society: American electroencephalographic society guidelines for standard electrode position nomenclature. J. Clin. Neurophysiol. **8**(2), 200–202 (1991). https://doi.org/10.1097/00004691-199104000-00007

11. Sanz, J., Wulff-Abramsson, A., Aguilar-Paredes, C., Bruni, L., Sánchez, L.: Synchronizing audio-visual film stimuli in unity (version 5.5. 1f1): Game Engines as a Tool for Research (2019). arXiv:1907.04926

12. Pfurtscheller, G., Neuper, C., Brunner, C., Lopes da Silva, F.: Beta rebound after different types of motor imagery in man. Neurosci. Lett. **378**(3), 156–159 (2005). https://doi.org/10.1016/j.neulet.2004.12.034

13. Pfurtscheller, G.: Functional brain imaging based on ERD/ERS. Vis. Res. **41**(10–11), 1257–1260 (2001). https://doi.org/10.1016/S0042-6989(00)00235-2

14. Pfurtscheller, G., Lopes Da Silva, F.: Event-related EEG/MEG synchronization and desynchronization: basic principles. Clin. Neurophysiol. **110**(11), 1842–1857 (1999). https://doi.org/10.1016/S1388-2457(99)00141-8

15. Lachaux, J.-P., Chavez, M., Lutz, A.: A simple measure of correlation across time, frequency and space between continuous brain signals. J. Neurosci. Methods **123**(2), 175–188 (2003). https://doi.org/10.1016/S0165-0270(02)00358-8

16. Heeren, T., D'Agostino, R.: Robustness of the two independent samples t-test when applied to ordinal scaled data. Stat. Med. **6**(1), 19–90 (1987). https://doi.org/10.1002/sim.4780060110

17. Neuhauser, M.: Nonparametric Statistical Tests: A Computational Approach. Chapman and Hall/CRC, Boca Raton (2011)

18. Drexler, W., Fujimoto, J.: Optical Coherence Tomography. Springer, Cham (2016). https://doi.org/10.1007/978-3-319-24817-2_8

19. Wilsgaard, T., Jacobsen, B.: Lifestyle factors and incident metabolic syndrome: the Tromsø study 1979–2001. Diabetes Res. Clin. Pract. **78**(2), 217–224 (2007). https://doi.org/10.1016/j.diabres.2007.03.006

20. Lafond, C., Series, F., Lemiere, C.: Impact of CPAP on asthmatic patients with obstructive sleep apnoea. Eur. Respir. J. **29**(2), 307–311 (2007). https://doi.org/10.1183/09031936.000 59706

21. Sims, R.: Bivariate Data Analysis: A Practical Guide. Nova Publishers, New York (2000)

22. Michels, L., Moazami-Goudarzi, M., Jeanmonod, D., Sarnthein, J.: EEG alpha distinguishes between cuneal and precuneal activation in working memory. Neuroimage **40**(3), 1296–1310 (2008). https://doi.org/10.1016/j.neuroimage.2007.12.048

23. Adam, A., Ibrahim, Z., Mokhtar, N., Shapiai, M.I., Mubin, M., Saad, I.: Feature selection using angle modulated simulated Kalman filter for peak classification of EEG signals. SpringerPlus **5**(1), 1–24 (2016). https://doi.org/10.1186/s40064-016-3277-z

24. Klimesch, W., Doppelmay, M., Pachinger, T., Ripper, B.: Theta band power in the human scalp an the encoding of new information. NeuroReport **7**(7), 9–12 (1996). https://doi.org/10.1097/00001756-199605170-00002

25. Doppelmayr, M., Klimesch, W., Pachinger, T., Ripper, B.: The functional significance of absolute power with respect to event-related desynchronization. Brain Topogr. **11**(2), 133–140 (1998). https://doi.org/10.1023/A:1022206622348

26. Gordon, E., Sim, M.: The EEG in presenile dementia. J. Neurol. Neurosurg. Psychiatry **30**(3), 285 (1967)

27. Rodriguez, E., George, N., Lachaux, J.-P., Martinerie, J., Renault, B., Varela, F.: Perception's shadow: long-distance synchronization of human brain activity. Nature **397**(6718), 430–433 (1999). https://doi.org/10.1038/17120

28. Reber, T., et al.: Intracranial EEG correlates of implicit relational inference within the hippocampus. Hippocampus **26**(1), 54–56 (2016). https://doi.org/10.1002/hipo.22490

29. Lyerly, S.: The average Spearman rank correlation coefficient. Psychometrika **17**(4), 421–428 (1952). https://doi.org/10.1007/BF02288917

30. Costa, T., Rognoni, E., Galati, D.: EEG phase synchronization during emotional response to positive and negative film stimuli. Neurosci. Lett. **406**(3), 159–164 (2006). https://doi.org/10.1016/j.neulet.2006.06.039

31. Krause, C.M., Viemerö, V., Rosenqvist, A., Sillanmäki, L., Aström, T.: Relative electroencephalographic desynchronization and synchronization in humans to emotional film content: an analysis of the 4–6, 6–8, 8–10 and 10–12 Hz frequency bands. Neurosci. Lett. **286**(1), 9–12 (2000). https://doi.org/10.1016/S0304-3940(00)01092-2

A Study of Colour Using Mindwave EEG Sensor

Ana Rita Teixeira[1,2](✉) and Anabela Gomes[3,4]

[1] IET - Institute of Electronics and Informatics Engineering of Aveiro, University of Aveiro, Aveiro, Portugal
ateixeira@ua.pt
[2] Coimbra Polytechnic - ESEC, Coimbra, Portugal
[3] Coimbra Polytechnic - ISEC, Coimbra, Portugal
anabela@isec.pt
[4] Centre for Informatics and Systems, University of Coimbra, Coimbra, Portugal

Abstract. Human-Computer Interaction (HCI) is a multidisciplinary research area aiming the design of user-friendly systems. Even though systems are increasingly complex, recurring more and more to multimodal interactions, there are very basic aspects, such as colour and its correct perception, that continue to be crucial for effective and pleasurable communication. These aspects will be reflected in several areas and may be particularly useful in teaching-learning systems.

Thus, we made an experiment designing a study allowing the identification of shapes, determined by the correct perception of certain colour combinations. Hence, we implemented a colour test similar to the Ishihara Colour Blindness Test, but instead of showing numbers it shows objects and several different shapes. In this experiment, we evaluated the users' responses and their feedback regarding the ease of determining the shapes in question, resulting from certain colour combinations. Additionally, we used an EEG signal to determine neurophysiological aspects that are impossible to manipulate with in relation to each displayed figure. From this, it was possible to determine aspects such as the levels of immersion, fatigue and stress caused by each combination.

Keywords: EEG · Mindwave · Fatigue · Immersion · Stress · Visual perception

1 Introduction

1.1 Contextualization

HCI is a growing and increasingly important area. This area is no longer just about designing user-friendly systems, but more and more attention is being given to the inclusion of innovative new forms of interaction and multimodal interactions. Although these increasingly innovative aspects, there are pertinent and fundamental features on which we consider crucial their true understanding. In this sense, we consider that the presentation of colours and their combination may affect considerably the users' perceptions.

Thus, we designed a study allowing the identification of shapes, determined by the correct perception of certain colour combinations, particularly important in multimodal interfaces where these basic aspects should represent the lowest possible cognitive load.

D. D. Schmorrow and C. M. Fidopiastis (Eds.): HCII 2020, LNAI 12196, pp. 176–188, 2020.
https://doi.org/10.1007/978-3-030-50353-6_13

Hence, we implemented a colour test similar to the Ishihara Colour Blindness Test [1], but instead of showing numbers it shows objects and several different shapes. There are 15 different levels, and in each level the participant should select one of the three answer options corresponding to the form that he/she sees. The test has a limit of 5 min. Although these types of tests are primarily used for the diagnosis of colour blindness, we believe that their use in non-colour-blind people may provide valuable insight into the recommendations to follow regarding certain colour combinations to be used at interfaces.

Colour is one of the most curious aspects of visual perception, so different colour perception could be found also in differences in brain activity. Colour combinations also influence legibility [2] and the inappropriate use of colours can result in higher levels of visual discomfort and poor reading performance [3]. Huang [4] found that visual search performance can be significantly affected by colour combinations. Therefore, we decided to use the electroencephalogram (EEG) signal to detect several parameters levels for the recognition of visual forms in an environment that is difficult to perceive. For that, we used Neurosky Mindwave headset, a portable device that was generally utilized to detect and measure electrical activity of the user's forehead. After the acquisition, the signal was studied in time and in frequency for extraction of parameters of interest. In this experiment, the idea is to confirm which combinations generate less fatigue, stress and immersion to provide faster and easier interactions freeing the user to more complex tasks. To compute fatigue, stress and immersion parameters the energy of important waves like Theta, Beta and Alpha was considered.

1.2 HCI and BCI

HCI is an expanding research area with lot of development in recent years. Most of the work in this area has as the concern the design of user-friendly systems and the most recent ones have as a concern the use of innovative interfaces such as voice, vision, gestures, virtual reality or augmented reality systems [5, 6]. Direct Brain-Computer Interface (BCI) adds a new set of possibilities to HCI [7].

BCI is a way of interaction between individuals and computers don't using any muscle, controlled through individuals' brain activity captured with specific equipment. For Wolpaw et al. [8], BCI is a communication system having two adaptive components complementing each other reciprocally. This author adds that a BCI is a communication system that allows an individual to convey her/his intention to the external world by purely thinking without depending on the brain's normal output channels of nerves and muscles [9].

Brain computer interfaces have been applied in several fields of research from medical [10–13], smart environments [14], neuromarketing and advertisement [15], educational [16, 17] and self-regulation, games and entertainment [18], and Security and authentication fields [19]. The recording of cortical neuronal activity can be done in different ways in BCI systems for instance through EEG (Electro-Encephalo-Graphy), where several electrodes are placed on the scalp. More and more researches use EEG to determine the cognitive load of visual information [20, 21].

Even though EEG has been used for diversified areas the present study used it to determine the visual perception of a certain form according to certain colour combinations. Although we used isolated figures the idea will be to extend these concepts for more complex applications and environments. The users are increasingly demanding and if the products they use are not intuitive and clear they easily abandon them. One of the basic aspects of any interaction lies in the good visual perception of the information, being the colours and their correct combination in their genesis. The colours affect the users' perception which may have an impact on the cognitive performance of tasks. Therefore, it is of great importance the obtaining, processing and feedback of visual information using EEG as a way to reinforce the recommendations for digital interface optimization design.

Hence, in this paper, we used BCI from the viewpoint of one aspect of multimedia interactions, the colour. We consider the colour mainly important for a good visual perception, however some authors also established the motivational factors related to colour and layout [22].

2   Methodology and Procedure

2.1   Main Goals

The main objective of this work was to investigate the effect of colour on brain dynamics in the analysis of complex figures, including different colours combinations to obtain forms. Thus, we are interested in understanding what happens in terms of the energy of the Theta, Beta and Alpha waves in each of the analysed figures, and consequently analyse the levels of Fatigue, Immersion and Stress in each one having an indication of the ease with which colour is perceived. Therefore, our study intended to address the following questions: Q1 - What are the levels and combinations that allow faster response times?; Q2 - Which levels and colour combinations have the highest levels of Fatigue, Immersion and Stress?

2.2   Experimental Design

The experiment was made in a usability laboratory, a calm and controlled environment in order to avoid interruptions. It consisted of a sequential visualization of fifteen squares having each one a form written in a certain combination of foreground and background colours not completely linear but consisting of dots of different sizes, colours and proximity distances.

There are 15 different figures, and in each one the person has to select one of the three answer options depending on what he/she sees (Fig. 1). Associated with each level, there will be a question to which the participants have to answer: "What is the shape you can see?". During this task, each figure was presented at the center of the screen along with the answer options. There is no time limit, for participants to respond. The answer consists of selecting the correct option, followed by the confirmation of that answer by a second click. After this, there is a pause with a white screen for 1 s and a new image appears (Fig. 2).

It should be noted that the location on the screen as well as the size of the images did not change. The participants were also asked not to blink and not to move their eyes, and the body during the visualization of the screens. The purpose of this procedure was

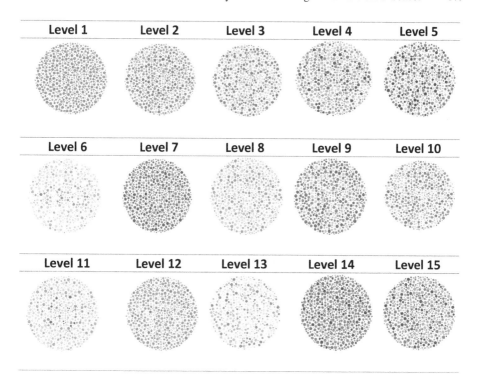

Fig. 1. Levels of the game

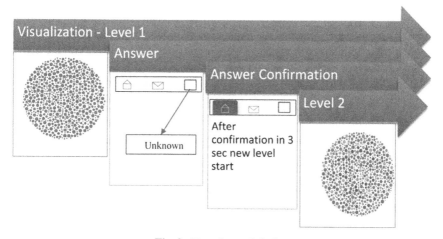

Fig. 2. Experimental design

to eliminate ocular and muscular artifacts, thus avoiding signal loss. At the end of the experiment, users were asked to classify in three groups of difficulty (Low, Medium, High), each one of the figures, according to the ease with which they discover the shapes.

2.3 Data Sample

Twenty-eight participants (8 males and 20 females), aged 18–22 yr (mean: 19 ± 0.8 yr), were recruited to perform the already mentioned task. All participants provided informed consent prior to participating in the study. At the beginning of the experiment, a questionnaire was made to each participant in order to obtain a more detailed characterization of that population: age, sex, level of education and the existence of visual problems. The purpose of the study was also explained to the participants.

2.4 Feature Acquisition

The feature acquisition was made using the Neurosky's Mindwave device for measuring the brain activity [23]. This simple and affordable device includes two electrodes, one for EEG dataset records (Fp1 channel), another for reference signals (the ear clip) and a power switch. The sample frequency is 512 Hz. The Mindwave EEG sensor processes the brainwave into digital signals and uses the eSense algorithm to compute user's engaging attention and concentration. The eeg_ID software was used to connect the Mindwave through Bluetooth. The EEG Raw Data can be recorded in a .csv file. In addition, the signals related to Alpha Low, Alpha High, Beta Low, Beta High, Delta, Gamma Low, Gamma High and Theta frequencies can be also recorded. Based on these information, three parameters were used in this study: Fatigue [24], Stress [25] and Immersion [26].

3 Methodology Results

3.1 Clustering Methods

To achieve the objectives of our study, we organized the data using 3 different clustering methods in order to verify whether the parameters (Fatigue, Stress and Immersion) were more or less discriminative. The clusters were prepared based on Error (average number of wrong answers for each level obtained by all participants), Time (average time spent on each level by all participants) and UX (User eXperience obtained through a survey made to students to classify each level in 3 degrees of difficulty (Low, Medium, High) (Table 1).

Table 1. Clustering levels considering error, time and user experience methods

Error method		Time method		UX method	
C1_E:	L9, L14	**C1_T:**	L9, L14	**C1_UX:**	L1, L12
C2_E:	L2, L3, L10, L11, L13, L15	**C2_T:**	L2, L3, L4, L5, L6, L7, L8, L10, L11, L12, L3, L15	**C2_UX:**	L2, L3, L4, L5, L6, L8, L10, L11, L13
C3_E:	L1, L4, L5, L6, L7, L8, L12	**C3_T:**	L1	**C3_UX:**	L7, L9, L14, L15

3.1.1   Error Method

Figure 3 shows the levels and the errors made in each of them by the totality of individuals. The clustering analysis led to the formation of the following clusters: C1, consisting of the levels L9 and L14 (where individuals made more mistakes); C2 consisting of the levels L2, L3, L10, L11, L13 and L15; C3 consisting of levels L1, L4, L5, L6, L7, L8 and L12 (where individuals made no mistakes).

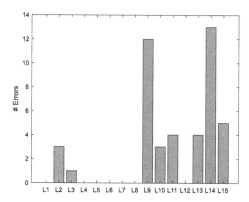

Fig. 3.   Number of errors in each level for all participants

3.1.2   Time Method

Figure 4 shows each of the levels and the response time obtained in each of them by the totality of individuals. The clustering analysis by response time led to the constitution of the following clusters: C1 consisting of levels L9 and L14 (where individuals had the longest response time); C2 consisting of levels L2, L3, L4, L5, L6, L7, L8, L10, L11, L12, L13 and L15; C3 constituted by L1 (where individuals had the shortest response time).

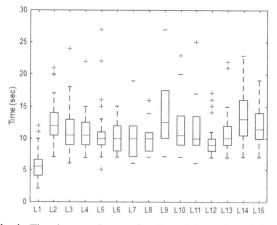

Fig. 4.   Time in seconds spent for all participants in each level

3.1.3 UX Method

Figure 5 shows each of the levels and the UX classification in each of them by the totality of individuals. The clustering analysis by the three defined difficulty levels (Low, Medium, High) was the following: C1 consisting of levels L1 and L12; C2 consisting of levels L2, L3, L4, L5, L6, L8, L10, L11 and L13 and C3 consisting of levels L7, L9, L14 and L15.

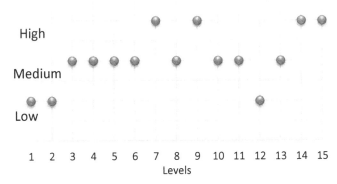

Fig. 5. Average of levels classification (Low, Medium and High) by all participants considering all levels

3.2 Features Analysis

Next the jointly boxplot for all the parameters analyzed, namely Immersion (Fig. 6), Fatigue (Fig. 7) and Stress (Fig. 8) are presented. The three different forms of data clustering were considered to compare the parameters values: Immersion, Fatigue and Stress. The main objective of this analysis is to understand the best clustering method in order to be able to discriminate the dataset. As can be seen by the figures, for all the parameters of interest, the most discriminative method is the UX method, so in the next section we will focus our analysis on this method of clustering.

Figure 8 shows that the Immersion parameter is more discriminative considering the UX and the Error methods in the data clustering. Note that for both methods there is a statistical difference between the values of each cluster and average terms considering a 90% confidence interval.

In the case of the Fatigue parameter, there are two methods that stand out, UX and Time, having the UX significant differences between all clusters (Fig. 7).

In the case of the Stress parameter, the differences between the different grouping methods are not so evident, although the UX method and the Error method can be highlighted as more discriminative.

From the general analysis, we concluded that the UX method stands out for all the considered parameters.

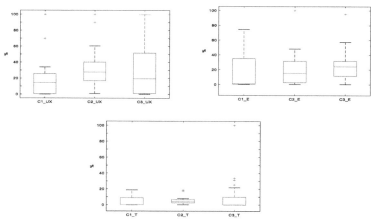

Fig. 6. Immersion values for each cluster (C1, C2 and C3) considering the three techniques of clustering: User Experience (UX), Time (T) and Error (E).

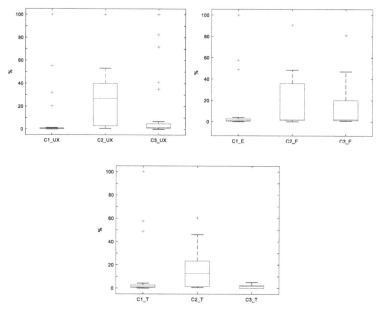

Fig. 7. Fatigue values for each cluster (C1, C2 and C3) considering the three techniques of clustering: User Experience (UX), Time (T) and Error (E).

3.3 Results Discussion

Next, the discussion of the results obtained for the organization of clustering, according to the User eXperince, will be presented. First, we observed that the average time spent in each cluster for all individuals was as follows: C1 of 7.72 s, C2 of 11.84 s and C3 of 12.48 s. We thus confirm that the lowest difficulty levels are found in cluster C1 and the

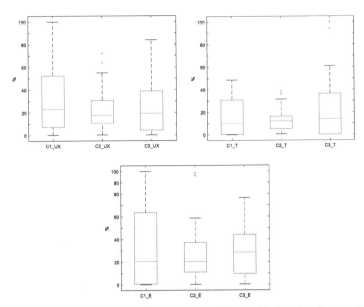

Fig. 8. Stress values for each cluster (C1, C2 and C3) considering the three techniques of clustering: User Experience (UX), Time (T) and Error (E).

most difficult levels are found in cluster C3. Calculating the T-test between each pair of clusters, it turns out that they are all different from each other, with the difference being the following C2 × C3 (H0 = 1, p = 0.0366, 90%); C2 × C1 (H0 = 1, p = 6.1 × 10–13, 90%); C3 × C1 (H0 = 1, p = 1.43 × 10–10, 90%).

Regarding Stress, the levels of Cluster C1 were, in all individuals, the ones that led to higher average levels (379.82), followed by those of cluster C3 (297.17) with the levels of cluster C2 obtaining the lowest average (291.94). The application of the T-test to ascertain the difference in means between each pair of clusters indicates that there is a difference between all, namely between C2 × C3 (H0 = 1, p = 0.62, 90%); C2 × C1 (H0 = 1, p = 0.14, 90%) and C3 × C1 (H0 = 1, p = 0.24, 90%).

Regarding Fatigue, the levels of Cluster C1 were those that, in all individuals, led to higher averages (9.51), followed by those of cluster C2 (6.50), with levels of cluster C3 obtaining the lowest average (4.08). The application of the T-test to ascertain the difference in means between each pair of clusters indicates that there is a difference between all, namely between C2 × C3 (H0 = 1, p = 0.19, 905); C2 × C1 (H0 = 1, p = 0.57, 90%) and C3 × C1 (H0 = 1, p = 0.29, 90%);

Regarding Immersion, the levels of Cluster C1 were, in all individuals, the ones that led to higher averages (165.91), followed by those of cluster C2 (101) with the levels of cluster C3 obtaining the lowest average (71.39). The application of the T-test to ascertain the difference in means between each pair of clusters indicates that there is a difference between all, namely between C2 × C3 (H0 = 1, p = 0.13, 90%); C2 × C1 (H0 = 1, p = 0.13, 90%) and C3 × C1 (H0 = 1, p = 0.0178, 90%).

The descriptive statistic of Fatigue, Immersion and Stress can be seen in Table 2, Table 3 and Table 4, respectively.

The results of the cluster analysis indicate that in cluster C1 the average response times are the lowest, meaning that it is the cluster that has the easiest levels to identify. An individual analysis shows that L1 is the one with the lowest response time, followed by

L12. These 2 levels were also correctly identified by all individuals. However, this is the cluster with the highest levels of fatigue, which means that these combinations in long periods will be avoided because they cause tiredness. The average level of immersion in this cluster is also the highest, with the same being verified in the individual analysis of each of these levels in relation to the totality of levels for all individuals, meaning combinations that lead to a good concentration.

Table 2. Clustering levels for fatigue, considering error, time and user experience method

Fatigue			
Min.	Med.	Max.	St. dv.
C1: 0.011	**C1:** 9.51	**C1:** 121.13	**C1:** 26
C2: 0.019	**C2:** 6.50	**C2:** 26.26	**C2:** 6.36
C3: 0.009	**C3:** 4.08	**C3:** 30.52	**C3:** 8.31

Table 3. Clustering levels for immersion, considering error, time and user experience method

Immersion			
Min.	Med.	Max.	St. dv.
C1: 0.36	**C1:** 165	**C1:** 954	**C1:** 215
C2: 2.63	**C2:** 103	**C2:** 319	**C2:** 75.65
C3: 0.15	**C3:** 71.39	**C3:** 255	**C3:** 17.78

Table 4. Clustering levels for stress, considering error, time and user experience method

Stress			
Min.	Med.	Max.	Std. dev
C1: 1.57	**C1:** 379.8	**C1:** 1262.8	**C1:** 341.3
C2: 3.86	**C2:** 291.9	**C2:** 912	**C2:**231.5
C3: 2.7	**C3:** 297.2	**C3:** 1060.2	**C3:** 278.7

In the C2 cluster, the average response times are higher than in the previous cluster, meaning that there are factors that make the levels more difficult to identify. In this cluster, although levels L4, L5, L6 and L8 were correctly identified by all individuals, these levels required more time to be correctly identified. The L4 is among the levels of this cluster with the highest average fatigue for all individuals, perhaps because it has a vibrant and clear background and writing colour (orange and green). According to Perron, certain contrasts create so much vibration that it diminishes readability [27, 28]. This level (L4) is similar, in general, to L2, having a combination of similar colours, but L4 has a less uniform distribution of colours in the background that can confuse the user. L5 is the level of this cluster that causes more fatigue (and the 2nd in all levels) and more immersion (and the 1st in all levels), perhaps because it has both a vibrant background and foreground colours (red and green) but is more attractive. L6 leads to less fatigue but some immersion. We can consider that this level contains the same colour combination as L4, but in a smoother and clearer way, especially with regards to the background colour. L8 despite having two light colours, they establish contrast generating high levels of immersion but also fatigue. L8 has a colour combination similar to L4 and L6 but inverted and less intense, taking a little more time to identify. L3 was not correctly identified by 1 individual and it is the level that has less immersion and the second that has less fatigue in this cluster. L3 has a less intense writing colour or with similarly coloured dots further away, taking longer to identify, but the response options may have given some clue towards the correct identification by a greater number of individuals. L2 and L10 were not correctly identified by 3 individuals, showing curiously low levels of fatigue in average terms for all individuals, but L2 shows high levels of immersion. L11 level was not correctly identified by 4 individuals, with relatively high levels of fatigue and immersion. The image shown in L11 is difficult to detect by moving away from similarly coloured dots. We can consider that L11 would have the same characteristics as L6, but the points of similar colour in the way to be detected are further apart having additionally larger points and less different colours nearby, making the task more difficult.

In the C3 cluster, response times are the highest. Levels L14 and L9 were the levels incorrectly identified by a greater number of individuals, respectively by 13 and 12 individuals as well as the levels with the highest level of fatigue in this group. Also, L15 was not correctly identified by 5 individuals. However, both this level and L7 obtained lower levels of fatigue within this cluster. In terms of immersion, and within this cluster, the levels obtained decreasing immersion values at the following levels L15, L7, L14 and L9. At the L9 level, colours with little contrast predominate and the figure to be detected has points of similar colour further away. L7 is similar to L9 but the colours are more intensified, becoming darker, perceiving better than L9 but causing more tiredness. In L14 orange over brown predominates, with little contrast. L9 and L15 are only perceived by the suggested answer.

4 Conclusions

The colour plays a key role in any graphic design of any interactive system. One of the main goals of its utilization is the readability and its correct perception in order for correct communication to occur. However, there are colour combinations that are more easily perceived and carry a message better than others while others produce visual discomfort, causing visual noise and sometimes even distorting the message that is communicated or, at least, tiring the user.

Despite growing knowledge about colour processing and understanding, it is not common to know how the brain processes the information in the presence of certain colour combinations. For that we made an experiment consisting in the identification of a certain shape in several figures having different colour combinations, using a non-conventional approach, the EEG signal. The idea was to collect several parameters not subjective or causing doubts but impossible to tamper with and getting information about the levels of immersion, stress or fatigue.

Although it was not possible to obtain a clear answer to the research questions, the study confirmed that a good chromatic contrast is essential to minimize the levels of fatigue and stress. In addition, it was found that the level of immersion can be caused either by contrasts that are difficult to perceive or by chromatic combinations that are attractive for the user.

We believe that this work represents a set of promising developments that could be very relevant in the area of HCI and with applications to many other areas.

References

1. Colblindor.: Ishihara 38 plates CVD test (2006–2018). http://www.color-blindness.com/ish ihara-38-plates-cvd-test/#prettyPhot. Accessed 2 Apr 2019
2. Nilsson, L.G., Ohlsson, K., Ronnberg, J.: Legibility of text as a function of color combinations and viewing distance. Umeå Psychol. Rep. **167** (1983)
3. Bruce, M., Foster, J.J.: The visibility of colored characters on colored background on view data displays. Vis. Lang. **16**, 382–390 (1982)
4. Huang, K.: Effects of computer icons and figure/background area ratios and color combinations on visual search performance on an LCD monitor. Displays **29**, 237–242 (2008)
5. Segen, J., Kumar, S.: Human-computer interaction using gesture recognition and 3D hand tracking. In: Proceedings of the IEEE International Conference on Image Processing, Chicago, IL, vol. 3, pp. 188–192 (1998)
6. Stiefelhagen, R., Yang, J.: Gaze tracking for multimodal human-computer interaction. In: Proceedings of the IEEE International Conference on Acoustics, Speech, Signal Processing, Munich, Germany, vol. 4, pp. 2617–2620 (1997)
7. Pfurtscheller, G., et al.: Current trends in graz brain-computer interface (BCI) research. IEEE Trans. Rehab. Eng. **8**, 216–219 (2000)
8. Wolpaw, J.R., Birbaumer, N., McFarland, D.J., Pfurtscheller, G., Vaughan, T.M.: Brain-computer interfaces for communication and control. Clin. Neurophysiol. **113**, 767–791 (2002)
9. Wolpaw, J.R., Birbaumer, N., McFarland, D.J., Pfurtscheller, G., Vaughan, T.M.: Brain-computer interfaces for communication and control. Clin. Neurophysiol. **113**(6), 767–791 (2020). https://doi.org/10.1016/S1388-2457(02)00057-3
10. Hanafiah, Z.M., Taib, M.N., Hamid, N.H.: EEG pattern of smokers for Theta, Alpha and Beta band frequencies. In: Proceedings of the 2010 IEEE Student Conference on Research and Development (SCOReD), pp. 320–323 (2010)
11. Padmashri, T.K., Sriraam, N.: EEG based detection of alcoholics using spectral entropy with neural network classifiers. In: Proceedings of the 2012 International Conference on Biomedical Engineering (ICoBE), pp. 89–93 (2012)
12. Sharanreddy, M., Kulkarni, P.: Detection of primary brain tumor present in EEG signal using wavelet transform and neural network. Int. J. Biol. Med. Res. **4**(1), 2855–2859 (2013)
13. Jones, C.L., Wang, F., Morrison, R., Sarkar, N., Kamper, D.G.: Design and development of the cable actuated finger exoskeleton for hand rehabilitation following stroke. IEEE/ASME Trans. Mechatron. **19**(1), 131–140 (2014)

14. Ou, C.-Z., Lin, B.-S., Chang, C.-J., Lin, C.-T.: Brain computer interface-based smart environmental control system. In: Proceedings of the 2012 Eighth International Conference on Intelligent Information Hiding and Multimedia Signal Processing, pp. 281–284 (2012)
15. Vecchiato, G., et al.: The study of brain activity during the observation of commercial advertisng by using high resolution EEG techniques. In: Proceedings of the 2009 Annual International Conference of the IEEE Engineering in Medicine and Biology Society, pp. 57–60 (2009)
16. Sorudeykin, K.A.: An educative brain-computer interface. Computer Science, Published in arXiv 2010, abs/1003.2660 (2010)
17. Shan, P., Pei, J.: Cognition and education management method of withdrawal reaction for students with internet addiction based on EEG signal analysis. Educ. Sci. Theor. Pract. 18(5) (2018). https://doi.org/10.12738/estp.2018.5.122
18. Tan, D.S., Nijholt, A.: Brain-Computer Interfaces: Applying Our Minds to Human-Computer Interaction. Springer, Heidelberg (2010). https://doi.org/10.1007/978-1-84996-272-8
19. Van de Laar, B., Gurkok, H., Plass-Oude Bos, D., Poel, M., Nijholt, A.: Experiencing BCI control in a popular computer game. IEEE Trans. Comput. Intell. AI Games 5(2), 176–184 (2013)
20. Khalifa, W., Salem, A., Roushdy, M., Revett, K.: A survey of EEG based user authentication schemes. In: Proceedings of the 8th International Conference on Informatics and Systems (INFOS). IEEE, pp. BIO–55 (2012)
21. Krigolson, O.E., Heinekey, H., Kent, C.M., Handy, T.C.: Cognitive load impacts error evaluation within medial-frontal cortex. Brain Res. 1430(1), 62–67 (2012). https://doi.org/10.1016/j.brainres.2011.10.028
22. Niu, Y., Xue, C., Li, X., Li, J., Wang, H., Jin, T.: Icon memory research under different time pressures and icon quantities based on event-related potential. J. Southeast Univ. (English Edn.) 30(1), 45–50 (2014). https://doi.org/10.3969/j.issn.1003-7985.2014.01.009
23. Bennett, K.B., Flach, J.: Display and Interface Design: Subtle Science, Exact Art. Design Principles: Visual Momentum. CRC Press, Boca Raton (2011)
24. NeuroSky MindWave User Guide (2018)
25. Jap, B.T., Lal, S., Fischer, P., Bekiaris, E.: Using EEG spectral components to assess algorithms for detecting fatigue. Expert Syst. Appl. 36, 2352–2359 (2009)
26. Quaedflieg, C., Meyer, T., Smulders, F., Smeets, T.: The functional role of individual-alpha based frontal asymmetry in stress responding. Biol. Psychol. 104, 75–81 (2015)
27. Lim, S., Yeo, M., Yoon, G.: Comparison between concentration and immersion based on eeg analysis. Sensors 19, 1669 (2019)
28. Perron, C.: Colour choices on web pages: contrast vs readability (2012). http://www.writer2001.com/index.htm

Ensemble Usage for Classification of EEG Signals A Review with Comparison

Zaib Unnisa[1]([envelope]), Sultan Zia[1], Umair Muneer Butt[1,2], Sukumar Letchmunan[2], and Sadaf Ilyas[1]

[1] University of Lahore, Gujrat 50700, Pakistan
zaib.unnisa82@yahoo.com,
{sultan.zia,umair.muneer}@cs.uol.edu.pk,
mirzasadafilyas@gmail.com
[2] University Sains Malaysia, Penang, Malaysia
sukumar@usm.my

Abstract. *Significance*: Ensemble learning is a robust and powerful approach to solve a variety of classification problems. Its usage has increased dramatically in recent years but not seen extensive application in EEG based Brain computer interface (BCI) problems. There is a wide range of classifiers which may not perform well, when used separately for classification problem but outperforms state-of-art algorithm when used as an ensemble, i.e., Long-short Term Memory (LSTM) is considered as best learning algorithms when time is embedded in input but not shown outstanding performance when used individually for EEG classification. On the contrary, it provided real good results when used as an ensemble. Aim: Aim of this study is how EEG signals can be classified using Ensembles methods; its importance and usage are described with experimental results. Approach: The approach that is being used is, combining different classifiers, i.e. Support vector machines, Decision Trees, Random Forest, Long Short Term Memory (LSTM), Logistic Regression (LR) and see which classifiers work best with which ensemble technique for EEG classification's problem. Datasets: Datasets are taken from well-known data resources, Kaggle, EEG data set of confused students. The second dataset is taken from GitHub having EEG signals with timestamps according to events, i.e., sound, light, etc. According to our results, the LSTM- ensemble outperformed all other algorithms in the case where time is embedded in data.

Keywords: Ensemble methods · Brain Computer Interface (BCI) · EEG signals · Classification

1 Introduction

Artificial Intelligence (AI) is used to bridge the gap between humans and machines. Brain Computer Interface (BCI) is going one step further by enabling the brain to operate machines and computers through brain waves. The human brain keeps on producing waves, which are the result of neural oscillations [1]. These brain waves influence physiologically, physically, emotionally i.e. all aspects of human life. These waves can be

© Springer Nature Switzerland AG 2020
D. D. Schmorrow and C. M. Fidopiastis (Eds.): HCII 2020, LNAI 12196, pp. 189–208, 2020.
https://doi.org/10.1007/978-3-030-50353-6_14

measured, visualized, and operated [2] therefore can be used for monitoring above mentioned aspects as well as can be used for controlling electrical devices and computers without any motor movement by using BCI [3]. BCI provides an alternative path to how humans can interact with devices without using their hands, feet or any other motor activity. There are different BCI systems. i.e. Invasive and Non-invasive, Synchronous and Asynchronous, Dependent and Independent. Datasets used in this study are collected by using non-invasive, synchronous and independent BCI systems that mean datasets are collected without any surgery, without considering any artifacts, i.e. eye movement or gaze, etc.

EEG signal's acquisition is safe and is used at a large scale for research purposes. There are different ways of measuring brainwaves i.e. intracranial electroencephalography (iEuEG) [4], Stereotactic Electro-encephalogram (sEEG) [4], Magneto-Encephalography(MEG) [5] and Electro-Encephalography(EEG) [6]. EEG based BCI is most prevalent in performing BCI tasks and used in a variety of applications, i.e., changing a TV channel [7, 8], controlling a wheelchair [9], home automation [10] so on. It has a wide range of applications from aiding Communication for Locked-in State (CLIS) patients to driving drones by healthy persons [11].

Researchers have been using different methods for correct and efficient classification of EEG signals. Various types of EEG signals are used for BCI; the most popular ones are P300 evoked potential, Steady State Evoked Potential (SSEV) and Motor Imagery (MI) EEG signals. P300 is an event-related potential (ERP) signal that is generated as a result of decision making. A person is asked to perform a sum, or, and, logical arithmetic task. While performing these tasks, signals are captured at the parietal lobe. The existence of these signals is used as a measure of cognitive functions [12]. SSEV signals are evoked as a result of visual stimuli. Suppression in Alpha waves is recorded by these waves [13]. Motor Imagery (MI) EEG signals are produced when a person "thinks" about doing any motor activity, without actually performing it. Brain waves that are responsible for doing the motor activity are produced in the motor cortex region of the brain. These MI signals are captured by placing electrodes on the motor cortex region of the brain. EEG signals are divided with respect to frequencies in these signals, Alpha, Beta, Theta, Gamma and Delta, as shown in Table 1. Decrease in Alpha waves, called Event-Related Desynchronization (ERD) and increase in beta waves, called Event-Related Synchronization (ERS) is recorded [11].

Table 1. Brainwaves with respect to frequency.

Frequency band	Frequency	States
Delta	0.5–4 Hz	Sleep
Theta	4–8 Hz	Deeply Relaxed
Alpha	8–12 Hz	Relaxed
Beta	12–35 Hz	Anxiety, Active
Gamma	35 Hz–greater	Concentration

Classification of EEG signals is a challenging task due to the involvement of eye movement and muscle movement; moreover, the amplitude of these signals is very low. Many spectral and spatial filters are used for performing pre-processing on the EEG signal before feature extraction [3]. It is followed by feature selection, to get rid of "curse of dimensionality". Different classifiers were used by researchers, e.g. Support Vector Machines (SVM), Multilayer Perceptron (MLP), K Nearest Neighbors (KNN) etc. in literature. Every method has its limits and none could prove its robustness like ensemble methods did [14].

Ensemble learning proved to be the best approach for solving broad areas of problems due to its flexibility, robust structure, and robustness. In this paper, different base classifiers are used separately, as well as in ensembles to compare their performance based on accuracy. Base classifiers which are focused in this study are, SVM, Decision trees, Long-short Term Memory (LSTM). Inspired by the extraordinary performance and efficiency of deep learning in most of the areas of classification problems, LSTM is the focal point of this study. LSTM methods have two motivational points; firstly, they are known as best when time is embedded in the input signal, which is the case in EEG signals. Secondly, deep learning's performance in literature encouraged to use this approach.

In the next section, the background study of ensemble methods as well as its types and significance is described. Various base classifiers, their merits, and demerits are presented. There is a brief discussion about weak learners. In the following section, methodology and structure are discussed, followed by experiments and results section, in which results are compared to conclude some solid points.

2 Background Study

In this section, past work described in the literature by various researchers is described.

2.1 Feature Extraction

EEG signals are non-stationary and adversely affected by noise [15]. Many methods have been used so far by researchers to eliminate unwanted signals. At first, a signal is acquired, and artifacts are removed. After that, features are extracted from the EEG signal. There are various feature extraction techniques discussed in the literature, as shown in Table 2.

Table 2. Feature extraction techniques

Method name	Advantages	Disadvantages	Analysis domain
FFT [16]	Good for narrowband signals i.e. sign waves and stationary signal processing	Not good at analyzing non-stationary signals i.e. EEG signals Suffers from Noise sensitivity	Frequency

(continued)

Table 2. (*continued*)

Method name	Advantages	Disadvantages	Analysis domain
Wavelet Transform [16]	Window size keeps changing Narrow at high frequency and broad at low frequency Suitable for a sudden signal change Useful for analyzing different data patterns. [15]	Needs a mother wavelet	Both time and frequency, Linear
Eigenvector [16]	Suitable for analyzing the sinusoid form of signals	May generate false zeros on low values	Frequency
Time-Frequency Distribution [16]	Suitable for analyzing continuous EEG signals	Needs noise-free signals for good results	Both time and frequency
Autoregressive [16]	Gives good frequency resolution Useful for analyzing short data segments	Too much dependency on the estimated model	Frequency
Independent Component Analysis (ICA) [17]	Preprocessing for removing artifacts Recover un-observed signals from given signals [18] ICA is better than PCA for ERP of EEG signals [18]	Can't determine variances of IC's	Time-frequency
Principal Component Analysis (PCA) [17]	Statistical Data Analysis Explain observed signals as linear combinations of orthogonal PCs PCA uses variance as a measure of the importance of vectors Not sensitive to scale effects [19] Aids in estimating probability in high dim of data Dramatically reduces data size	Assume methods (linear) Too expensive for some applications	Time and Frequency

PCA seeks direction which best represents the data, where an ICA finds directions that are independent of each other. ICA is often used in the tome series separation. From the above comparison, it is evident that wavelet and ICA are suitable for feature extraction from EEG signals.

2.2 Feature Selection

For EEG signals sometimes channels are selected instead of features, i.e., when classifying motor imagery tasks, then selected channels are C3, C4, and Cz [20]. Selecting frequency bands can reduce the task of feature selection. Algorithms are also used for selecting useful channels, i.e. spatial filtering, Common Spatial Filtering (CSP).

2.3 Classification

There are many classification approaches used so far for EEG classification but in this study, ensemble methods are focused.

Ensemble Methods: Ensemble methods are the most powerful tool for solving many classification problems. In this section, ensemble methods are described as well as some famous base classifiers are discussed.

Bagging: Bagging is anonymous of "**B**ootstrap **Aggreg**ating". For data training, it uses the bootstrap technique. One classifier learns on a subset of data and learns a rule. The second classifier learns on another subset of data, and the process continues for the nth classifier. This subset of data is chosen with replacement. For a classification problem, the output is generated by majority voting. A class about which most of the classifiers vote will be chosen as predicted class: for the regression problem, the output is predicted by the averaging output of all classifiers. Ensemble methods reduce the risk of overfitting, as prediction is not affected by individual data points [21]. Other than averaging and majority votes, there are some different strategies for predicting output i.e., Minimum, Maximum, Product, Sum, Decision Template, Bayes, Depster Shefer Fusion [22].

Boosting: Adaboost, which is known as boosting, is a bit different from bagging in a way that, instead of picking up data points at random, data points are selected according to their weights. These weights are assigned to data points as per their classification i.e. if a data point is classified correctly, then associated weights will be reduced; if the data point is misclassified, then its weight will be increased. By increasing weights of data points, their probability of getting selected by the next classifier will be increased. Therefore, the ensemble performs much better than the individual classifier by repeatedly training on "hard" examples. Hard examples are ones that are difficult to classify [21].

Algorithm 1:[21]

> - Convert Training $\{(X_t,y_t)\}$, $y_t \approx \{-1,1\}$
> - For t = 1 to T
>   - Construct W_t.
>   - Find Weak Classifier $h_t(x)$ with small error.
>   - $\epsilon_t = Pr_{W_t} [h_t(x) \neq y_t]$
> - Output H_{final}.

Suppose there is training data (X_i, Y_i) as shown in algorithm 1, where X_i is training instance and Y_i is the target value. The target value will be either $+1$ or -1. This loop will continue for T time steps. With weights, W for each time step weak classifier $h_t(x)$ is found that classifies at least more than chance training instances correctly. This classifier $h_t(x)$ will have error ϵ_t, which can have a value of less than half. Initially, equal weights are assigned to training examples; after each time step, weights are updated by using Eq. 1 [21].

$$Wt + 1 = \left(W_t(i). \, e^{-\alpha t. \, yi. \, ht(x)}\right) / Z_t \tag{1}$$

Where

$$\alpha_t = \frac{1}{2} \ln (1 - \epsilon_t / \epsilon_t)$$

Z_t is a normalizing factor, and α is the learning rate, which a positive constant value. The final hypothesis can be calculated by Eq. (2).

$$H_{final}(x) = sgn \left(\sum \alpha_t \cdot h_t(x) \right) \tag{2}$$

Weak Learners. In boosting examples, more importance is given to the examples which possess more weight. Weight shows how much emphasis should be given to this example. Weights are assigned to examples to make them "correctly" classified. A weak learner is one that correctly classifies samples more than chance (more than half) irrespective of weights that are assigned to examples. An ensemble is made of a combination of weak learners. A weak learner who cannot classify more than chance (50%) is not considered as a weak learner.

Gradient Boosting. Gradient boosting is an extension of the boosting algorithm. It is described by Eq. (3).

$$Gradient\ Boosting = Gradient\ descent + Boosting \tag{3}$$

Gradient Boosting can optimize any differentiable loss function. It may undergo overfitting. To avoid overfitting careful tuning hyper-parameters is needed [23].

Random Sub Space: Random Subspace is proposed by Ho [24]. It is similar to bagging except in this method features (attributes) of data points are randomly sampled with a replacement for each of the learners [24]. It is the right choice where the number of features is much larger than the training data points. It is applied to random forest, support vector machine, and nearest neighbor learners [25].

Base Classifiers: In this Section, some famous base classifiers are described.

Decision Tree (DT). By combining decision trees, better predictions can be made. A decision tree has low bias and high variance. By combining decision trees as an ensemble, low variance and low bias can be attained. Bagging is typically used for decision trees [26]. Random Forest is an ensemble method used for decision trees, and it is a trendy method. It de-correlate decision trees to a small extent.

Support Vector Machines (SVM). A support vector machine is used for the classification of data. it is done by making a hyperplane or set of hyperplanes in infinite-dimensional space. Hyperplane, which has the most significant margin to its supporting vectors, is considered as an optimal one [22]. A more considerable margin means a low generalization error of the classifier. SVM is a binary classifier. For nonlinear data separation, data is transformed into higher dimensions before applying SVM [22].

K Nearest Neighbor (KNN). This algorithm assumes that all the data points which are closer to a center point belong to the same class [22]. Data points which are closer to other central points will belong to different category. Distance learning algorithms are used for computing distance between data points and their central point. There will be total k primary points in sample space. This algorithm is based on instance-based learning. This algorithm is famous for solving classification and pattern recognition problems [22]. When a new data point is presented, its distance from all the k points is calculated and based on similarity measure; class is assigned to the unique data point.

Multilayer Perceptron (MLP). It is a Feed-Forward Neural Network (FFNN). MLP has one or more hidden layers, having neurons. Sigmoid or hyperbolic tangent are popular choices for activation functions for these neurons. These neurons are connected with weighted connections [22]. MLP is a popular choice where linear classifiers cannot solve the problem.

Long Short-Term Memory (LSTM). LSTM is the architecture of the Recurrent Neural Network. It is used in deep learning. It is famous for solving problems where time is embedded in the input. Many recent studies are showing the successful use of LSTM in different fields. To use previous predictions on the current tasks and keep track of forecasts so it can be used in the future, this is done in LSTM. It is capable of learning long term dependencies [27]. Long term memory is added in LSTM to remember previous predictions. It works, as shown in Fig. 1. Combinations of new information and prior forecasts are used as input. When a prediction is made, that is also saved in memory for use next time. Some of the predictions are forgotten, so memory may not accumulate with unnecessary information. For forgetting useless information, a neural network is used as shown in Fig. 1. as a sigmoid function. There is a selection gate, which is used to select which predictions should be output and which should not.

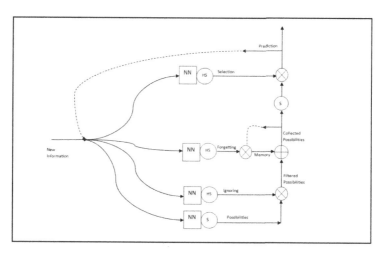

Fig. 1. LSTM architecture [28]

Whenever new information comes, lots of possibilities are generated. Only those possibilities are selected, which are relevant, and the rest are ignored. These possibilities are combined with the previous predictions in the memory; as a result, a lot of predictions are made. The selection gate selects which prediction should be sent on output [27, 28]. It is very beneficial practically i.e., in language translation, speech to text translation for any data that is embedded in time audio, video, in robotics or any other signals where the timestamp is used, or track of the previous history should be kept.

3   Methodology

In this section, those ensemble methods are discussed that are used for the classification of EEG signals by various authors in literature. If an ensemble of one classifier is used as a base classifier, it is called a homogenous ensemble method. If different classifiers are used as base classifiers, it is called a heterogeneous combination. Accuracy is determined by using this formula

$$\text{Accuracy} = \frac{(TN + TP)}{(TN + TP + FP + FN)} \quad (4)$$

TP shows true positive, i.e. the number of data points/training examples correctly classified. TN shows true negative, i.e., no. of data points that are not a member of concept/target class and "correctly" not classified as part of the target class. FN shows False negative, where algorithm classifies positive example as negative. FP is the stage where the algorithm detects a negative example as positive [15].

Table 3. Ensembles methods used in literature for EEG signals

Author and year	Feature extraction and selection	Ensemble methods	Base classifiers	Accuracy
Hosseinia en et al. (2018) [26]	Wavelet/ICA and Infinite ICA	Random Subspace	SVM	97%
Ali Al-Taei1(2017) [29]		Bagging	RF, Kstar	97.27%
Khalid et al. (2015) [30]	DWT and Four Statistical measures	Boosting	ANN, NB, K-NN, SVM	90%
Saugat et al. (2014) [31]	AAR and Estimated AAR coefficients by Least mean Square	Adaboost, LP Boost, RUS Boost, Bagging, RS	FFNN	88.57%, 81.90%, 60%, 75.7%, 75.71%
Jean et al. (2018).	Wavelet and LR with Elastic Net penalty	Boosting	Weighted LR	81.3%

(continued)

Table 3. (*continued*)

Author and year	Feature extraction and selection	Ensemble methods	Base classifiers	Accuracy
Hosseini et al. (2018) [32]	I-ICA	Random Sub Space	SVM, MLP	97%
Raviraj et al. (2018) [33]	CNN	Boosting	ConvLSTM	89.87%
Chun et al. (2014) [34]	NWFE	Majority Voting, Bagging	GC, SVM, RBFNN	91.6%
Kamlesh et al. (2014) [15]	Wavelet Transform and DWPT	Bagging	ANN, K-NN,SVM	88.7%
Ryan et al. (2017) [35]	Wavelet Transform	Bagging	SVM-L, RBF, SVM-R, FFNN, LSTM	83.6, 84.6, 84.4, 91.1, 91.8, 93.0

By comparing different ensemble methods in Table 3, it is pretty apparent that Bagging with random forest proved useful for the EEG classification problem. Figure 2 describes the percentage of usage of ensemble methods in the literature for the classification of EEG signals.

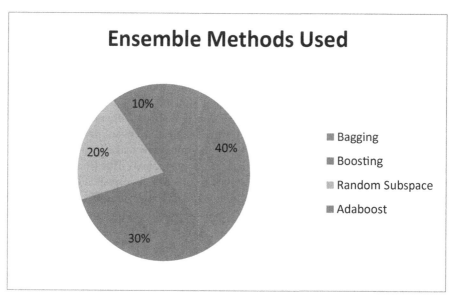

Fig. 2. Ensemble methods used in the literature review

This figure describes that bagging is the most popular method used in research in the case of the classification of EEG signals (Fig. 3).

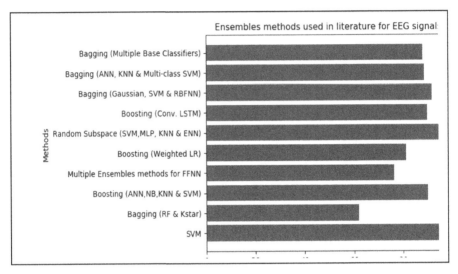

Fig. 3. Ensembles methods used in literature for EEG signals

The above figure shows the comparison of accuracies different ensemble methods used in literature. It shows Ensemble with SVM as a base classifier and Random Subspace are providing promising results.

4   Experiments and Results

In this study, the first data set is taken from a well-known source Kaggle, for the confused student. This data set is collected from 10 college students. Students are asked to watch two types of videos; first ones have some basic concepts about mathematics and algebra. The second one has challenging ideas with which students were not familiar with [36]. Every video was 2 min long and chopped in the middle to make it more confusing. Students wore a mindset having one electrode that is positioned on the frontal lobe. After each session, student rates their confusion state from 1 to 7, 1 corresponds to minimum confusing, and 7 corresponds to a maximum confusing state. Dataset is divided into columns, as shown in Table 4. The frequency band, which is playing a critical role in determining the user's confusion level, is theta.

Table 4. Confused student dataset

Column Number	Column names	Description
1	Subject ID	Student's ID
2	Video ID	Out of 20 which video student is watching
3	Attention	Property measure of mental focus
4	Meditation	Property measure of mental calmness
5	Raw	Raw EEG signal
6	Delta	1–3 Hz
7	Theta	4–7 Hz
8	Alpha 1	8–11 Hz
9	Alpha 2	8–11 Hz
10	Beta 1	12–29 Hz
11	Beta 2	12–29 Hz
12	Gamma 1	30–100 Hz
13	Gamma 2	30–100 Hz
14	Predefined Label	Whether the subject is expected to be confused
15	User-Defined Label	Whether the subject is confused

The second dataset is taken from Github; It is a multi-class classification problem; the EEG signal needs to be divided into 6 classes based on six events. Data is acquired from 32 channel electrodes; there are in total of 34 columns and 3469302 rows. There are 6 event types. Each sample will run for 1300 ms. In this dataset, time is embedded with input data. Data is preprocessed to remove Null values.

Experiment 1. Sklearn library is used in python 3.6 for data preprocessing in this study. Keras library and tensor flow are used at the backend. Columns such as Predefined-label, subject Id, Video Id, Attention, and Meditation are dropped. Any rows which have null values are also removed. In the next step, data is split into training and testing. 70% of data is used for training and 30% is used for testing. As the target value, the user's defined label is chosen. First of all, the decision tree classifier is applied, which gave 61% accuracy. After that, a random forest is used, which is an ensemble of decision tree classifiers. The value of the parameter "n_estimator" is set as 80, which shows there will be 80 decision trees in a random forest.67.9% accuracy is achieved, i.e. a 6.9% increase in accuracy by using ensemble. Afterward, bagging (Bootstrap aggregating)is applied to the same data set by using a decision tree as the base classifier. It scored 67.84% accuracy. In the next step boosting is used with the learning rate is .30 and n_estimator = 80. SVM is the base classifier is also used for Bagging and attained an accuracy of 50.2%, as shown in Table 5. Also used K nearest neighbor as a base classifier for bagging and used gradient boosting with learning rate as .30 and n-estimator equals to 100. The accuracy that is attained by gradient boosting is 60.8% as shown in Table 5.

The same experiments are repeated on the second dataset, with the same values of hyper parameters.

Table 5. Accuracies of different ensemble methods with different base classifiers.

Model	Base classifier	Accuracy	
		Dataset 1	Dataset 2
Decision Tree	–	60%	62%
Ensemble – Random Forest	Decision Tree	67.9%	74.2%
Bagging	Decision Tree	67.84%	75.1%
Boosting (AdaBoost)	Decision Tree	60.7%	74.3%
Bagging	SVM	50.2%	69.3%
Bagging	LDA	63%	67.8%
Bagging	KNN	57.25%	68%
Gradient Boosting	Decision Trees	60.8%	69.9%

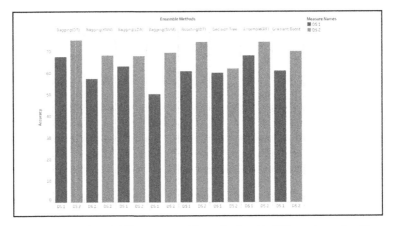

Fig. 4. Accuracies of different ensemble methods with different base classifiers

Figure 4 is showing the comparison of accuracies of different ensemble methods used in experiment 1 by both data sets. It is evident from the above results that bagging with decision trees and ensemble methods by using random forest proved better in comparison with others.

Experiment 2. Multilayer Perceptron (MLP) is used with hyper parameters, activation function "Relu", learning rate equals to 0.0001, 100 hidden layers, optimizer as "adam". Its accuracy is measured by computing confusion matrix, as shown in the following table (Table 6).

Table 6. Confusion Matrix

N	Predicted No	Predicted Yes	
Actual No:	TN (750)	FP (1160)	TN + FP
Actual Yes:	FN(671)	TP (1263)	FN + TP
	TN + FN	FP + TP	

Accuracy of MLP is (TP + TN)/n = (750 + 1263)/3844 = 52.36%

Performance Measures

TN: Predicted No and actual value is No too.

TP: Predicted Yes, and the actual value is Yes too.

FN: Predicted No, but the actual value is Yes.

FP: Predicted Yes, but the actual value is No.

Performance measures can be calculated by using formulae, as shown in Table 7.

Table 7. Formulae for calculating performance measures

Performance measure	Formula	Description
Accuracy	(TP + TN)/Total (n)	How often classifier is correct
Error rate	(FP + FN)/Total (n)	How often classifier is wrong
True Positive **Recall**	TP/Actual Yes	How many times it says yes when there is actually yes
False Positive	FP/Actual no	How many times it says yes when there is no
True Negative	TN/Actual No	How often it says no when there is no
Precision	TP/Predicted Yes	How often it says yes when there is actual Yes
Prevalence	Actual Yes/Total (n)	How often yes condition occurs

In the next step, a heterogeneous ensemble is used, in which three base classifiers are applied, Logistic Regression (LR), Random Forest (RF), and Support Vector Machine (SVM). In this ensemble, the vote cast approach is used. Accuracy measures are shown in Table 8.

The above table shows Random Forest is performing better than Logistic Regression and Support Vector Machines to a reasonable extent when used separately. Voting Classifier (Ensemble) is not proved useful for the classification of EEG signals. Figure 5 can view a graphical representation of this table.

Table 8. Vote casting Classifier

Model	Classifier	Accuracy	
		Dataset 1	Dataset 2
Logistic Regression	LR	52.315%	67.12%
Random Forest	RF	62.90%	73.6%
Support Vector Machines	SVM	50.31%	69%
Voting Classifier – Ensemble	All the above	53%	69%

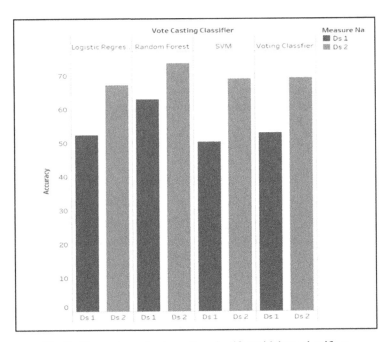

Fig. 5. Comparison of vote casting classifier with base classifiers.

In Fig. 5. The accuracy of the vote casting classifier compared with the base classifiers used. It is observed that the random forest attained higher accuracy than the vote casting ensemble method.

Experiment 3: A sequential model is defined in experiment 3. Neural Network is applied on the same dataset with accuracy 51.6%. Logistic function "sigmoid" is used as an activation function and adam optimizer to reduce loss. Results can be seen from the following Fig. 6.

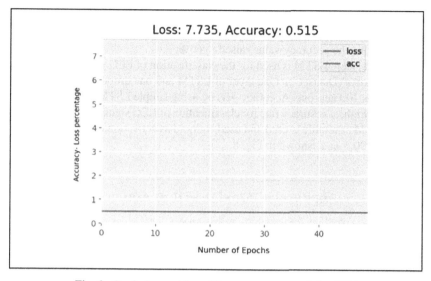

Fig. 6. Confusion matrix and Loss-accuracy graph for ANN

From the above result, it is evident that accuracy is not attained up toa satisfactory level. In the next step, more layers are added to the network. Activation functions used are, tanh and sigmoid. By increasing the number of layers, accuracy increases to 52%, as shown in Fig. 7.

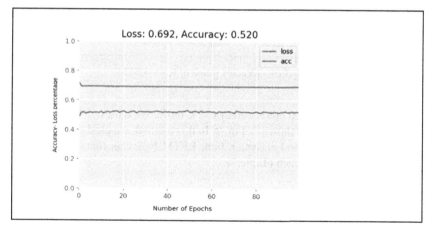

Fig. 7. Loss-accuracy graph after adding 3 more layers

Adding more layers didn't improve accuracy. This shows the Neural network is not working well-performing on the EEG classification's problem. By adding more layers in NN, overfitting also occurs. That's why in the next step, Recurrent Neural Network (RNN) is used.

Experiment 4: In the first step, LSTM is used on Dataset 1, in which time is not part of input data. LSTM with 100 neurons gave an accuracy of 58%. When LSTM is used with an ensemble, the accuracy value raised to 65%.

In the second step, LSTM is used for the classification of EEG signals in which time is part of input data (Dataset 2). One layer of LSTM and one dropout layer is used. The LSTM layer has 100 neurons. Accuracy gets by using simple LSTM is 78.6%, which is, in our case, the highest accuracy rate for classification of EEG signals with one classifier. In the next LSTM ensemble is used and used sigmoid as activation function and attained an accuracy of 90.3%, as shown in Fig. 8.

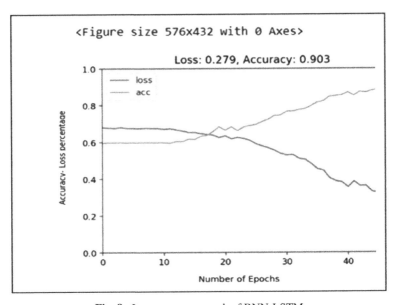

Fig. 8. Loss-accuracy graph of RNN-LSTM

LSTM ensemble improved this accuracy to 7% more. We achieved **97.8% accuracy** by using LSTM ensemble as shown in Fig. 9. In this method, we used the output of one classifier as input to the second classifier. Four LSTM base classifiers are used in this homogenous ensemble approach (Table 9).

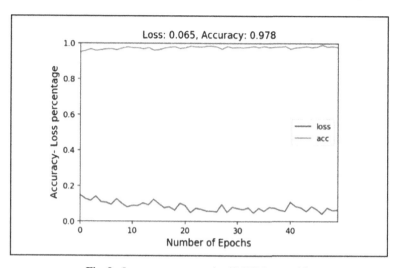

Fig. 9. Loss-accuracy graph of LSTM ensemble

Table 9. Long Short Term Memory-RNN Classifier

Model	Classifier	Accuracy	
		Dataset 1	Dataset 2
Long-Short Term Memory-RNN	LSTM	58.2%	78.6%
Bagging	LSTM	65.1%	90.3%
Boosting	LSTM	67.3%	**97.8%**

From the above experiments as results, we concluded that LSTM outperformed all other methods while the classification of EEG signals, by securing accuracy 97.8%. That shown LSTM ensemble is the best approach for EEG classification (Fig. 10).

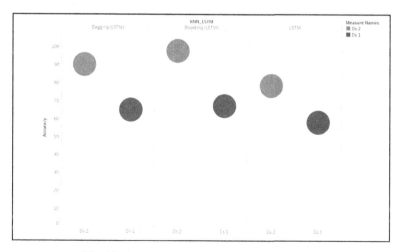

Fig. 10. Loss-accuracy graph of LSTM Ensemble

Figure 11 shows a comparison of all the classifiers used in this study.

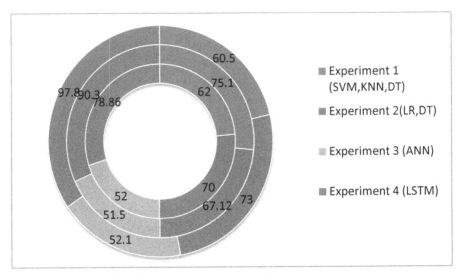

Fig. 11. Comparison of results of four experiments.

The first circle shows the application of a single method or one base classifier. The second circle depicts the results of bagging in the case of four experiments. The outer circle shows accuracies attained by applying by boosting. Figure 11 clearly shows the LSTM ensemble method achieved the highest accuracy with boosting ensemble method.

5 Conclusion

In this study, many classifiers are separately as well as in ensembles. From experiments, it is observed Boosting, and homogenous approach seems to have much more accuracy than other techniques of the ensemble. Moreover, LSTM outperformed as compared to state-of-the-art algorithms for the classification of EEG signals as per the results of our experiments with accuracy 90.3%. LSTM ensemble with a boosting approach further improved this accuracy to 7.5%, i.e., **97.8%** accuracy is achieved when time is embedded in the input signal (dataset 2). By reviewing the literature, it is observed that Bagging is giving promising results with Random Forest as a base classifier, but with experiments in this study, it is concluded that boosting performs better with LSTM as the base classifier. In future work, the LSTM ensemble will be used for the training classification model of MI EEG signals with self-acquired datasets. It will be used primarily in scenarios where MI EEG signals consist of low ERD/ERS values. It will be used in real-time situations for performing BCI applications.

References

1. Mureşan, R., Jurjuţ, O., Moca, V., Singer, W., Nikolić, D.: The oscillation score: an efficient method for estimating oscillation strength in neuronal activity. J. Neurophysiol. **99**, 1333–1353 (2008). https://doi.org/10.1152/jn.00772.2007
2. Wolpaw, J., Wolpaw, E.W.: Brain-Computer Interfaces Principles and Practice. Oxford University Press, Oxford (2012)
3. Lotte, F., et al.: A review of classification algorithms for EEG-based brain-computer interfaces: a 10-year update. J. Neural Eng. 0–20 (2018). https://doi.org/10.1088/1741-2552/aab2f2
4. Brain Waves - an overview| ScienceDirect Topics, https://www.sciencedirect.com/topics/agricultural-and-biological-sciences/brain-waves
5. McClay, W.A., Yadav, N., Ozbek, Y., Haas, A., Attias, H.T., Nagarajan, S.S.: A real-time magnetoencephalography brain-computer interface using interactive 3D visualization and the hadoop ecosystem. Brain Sci. **5**, 419–440 (2015). https://doi.org/10.3390/brainsci5040419
6. Haas, L.F.: Hans Berger (1873-1941), Richard Caton (1842–1926), and electroencephalography. J. Neurol. Neurosurg. Psychiatry **74**, 9 (2003). https://doi.org/10.1136/jnnp.74.1.9
7. Samsung is working on a TV that you can control using your thoughts. https://www.telegraph.co.uk/technology/2018/11/13/samsung-working-tv-can-control-using-thoughts/
8. I 'think' I'll change the channel. http://developer.neurosky.com/features/channel-change/
9. Meggiolaro, M.A.: Implementation of a wheelchair control using a four. Command Brain Comput. Interface **6**, 120–131 (2014)
10. Sharma, V., Sharma, A.: Review on: smart home for disabled using brain. Comput. Interfaces **2**, 142–146 (2015)
11. Tabar, Y.R., Halici, U.: A novel deep learning approach for classification of EEG motor imagery signals. J. Neural Eng. **14**, 16003 (2017). https://doi.org/10.1088/1741-2560/14/1/016003
12. van Dinteren, R., Arns, M., Jongsma, M.L.A., Kessels, R.P.C.: P300 development across the lifespan: a systematic review and meta-analysis. PLoS ONE **9**, e87347 (2014). https://doi.org/10.1371/journal.pone.0087347
13. Norcia, A.M., Appelbaum, L.G., Ales, J.M., Cottereau, B.R., Rossion, B.: The steady-state visual evoked potential in vision research: a review. J. Vis. **15**, 4 (2015). https://doi.org/10.1167/15.6.4
14. Datta, A., Chatterjee, R.: Comparative Study of Different Ensemble Compositions in EEG Signal Classification Problem. Presented at the (2019)
15. Kanoje, B.K., Shingare, A.S.: Automatic sleep stage detection of an EEG signal using an ensemble method. Int. J. Adv. Res. Comput. Eng. Technol. **8**, 2717–2724 (2014)
16. Al-Fahoum, A.S., Al-Fraihat, A.A.: Methods of EEG signal features extraction using linear analysis in frequency and time-frequency domains. ISRN Neurosci. **2014**(730218) (2014)
17. Singh, B., Wagatsuma, H.: A removal of eye movement and blink artifacts from EEG data using morphological component analysis. Comput. Math. Methods Med. **2017**, 1–17 (2017). https://doi.org/10.1155/2017/1861645
18. Bugli, C., Lambert, P.: Comparison between principal component analysis and independent component analysis in electroencephalograms modelling. Biometr. J. **49**, 312–327 (2007). https://doi.org/10.1002/bimj.200510285
19. Feature selection with mutual information, Part 2: PCA disadvantages. http://www.simafore.com/blog/bid/105347/Feature-selection-with-mutual-information-Part-2-PCA-disadvantages
20. Lotte, F.: A Tutorial on EEG Signal Processing Techniques for Mental State Recognition in Brain-Computer Interfaces. Eduardo Reck Miranda; Julien Castet. Guide to Brain-Computer Music Interfacing (2014)

21. Machine Learning – Udacity. https://classroom.udacity.com/courses/ud262/lessons/367378 584/concepts/3675486220923

22. Hussain, L., Aziz, W., Khan, A.S., Abbasi, A.Q., Kazmi, Z.H., Abbasi, M.M.: Classification of electroencephalography (EEG) alcoholic and control subjects using machine learning ensemble methods. J. Multidiscip. Eng. **2**(1), 126--131 (2015)

23. Decision Tree Ensembles- Bagging and Boosting – Towards Data Science. https://towardsdatascience.com/decision-tree-ensembles-bagging-and-boosting-266a8ba60fd9

24. The random subspace method for constructing decision forests: Tin Kam Ho, T.K. IEEE Trans. Pattern Anal. Mach. Intell. **20**, 832–844 (1998). https://doi.org/10.1109/34.709601

25. Sun, S., Zhang, C., Zhang, D.: An experimental evaluation of ensemble methods for EEG signal classification. Pattern Recognit. Lett. **28**, 2157–2163 (2007). https://doi.org/10.1016/j.patrec.2007.06.018

26. Decision Tree Ensemble Methods – Russell – Medium. https://medium.com/@rnbrown/decision-tree-ensemble-methods-6a89181b7083

27. Understanding LSTM Networks – colah's blog. http://colah.github.io/posts/2015-08-Understanding-LSTMs/

28. How recurrent neural networks (RNNs) and long-short-term memory (LSTM) work| End-to-End Machine. https://end-to-end-machine-learning.teachable.com/courses/516029/lectures/9533961

29. Al-Taei, A.: Ensemble Classifier for Eye State Classification using EEG Signals (2017). arXiv:1709.08590

30. Abualsaud, K., Mahmuddin, M., Saleh, M., Mohamed, A.: Ensemble classifier for epileptic seizure detection for imperfect eeg data. Sci. World J. **2015**, 1–15 (2015). https://doi.org/10.1155/2015/945689

31. Bhattacharyya, S., Konar, A., Tibarewala, D.N., Khasnobish, A., Janarthanan, R.: Performance analysis of ensemble methods for multi-class classification of motor imagery EEG signal, 712–716 (2014)

32. Hosseini, M.-P., Pompili, D., Elisevich, K., Soltanian-Zadeh, H.: Random ensemble learning for EEG classification. Artif. Intell. Med. **84**, 146–158 (2018). https://doi.org/10.1016/j.artmed.2017.12.004

33. Method, E., Conference, I.: Intelligent human computer interaction. 10688 (2017). https://doi.org/10.1007/978-3-319-72038-8

34. Chuang, C.H., Ko, L.W., Lin, Y.P., Jung, T.P., Lin, C.T.: Independent component ensemble of EEG for brain-computer interface. IEEE Trans. Neural Syst. Rehabil. Eng. **22**, 230–238 (2014). https://doi.org/10.1109/TNSRE.2013.2293139

35. Hefron, R.G., Borghetti, B.J., Christensen, J.C., Kabban, C.M.S.: Deep long short-term memory structures model temporal dependencies improving cognitive workload estimation. Pattern Recognit. Lett. **94**, 96–104 (2017). https://doi.org/10.1016/J.PATREC.2017.05.020

36. Wang, H.: Confused student EEG brainwave data | Kaggle. https://www.kaggle.com/wanghaohan/confused-eeg

A Reliable Neurophysiological Assessment of Stress – Basic Foundations for a Portable BCI Solution

Thomas Zoëga Ramsøy[1]([⊠]), Alexander Lopera[1], Philip Michaelsen[1], Mike Storm[1], and Ulrich Kirk[2]

[1] Neurons Inc., Taastrup, Denmark
thomas@neuronsinc.com
[2] University of Southern Denmark, Odense, Denmark

1 Introduction

Stress is among the most disturbing yet most prevalent mental disorders in our modern world. Besides the psychological impact that stress has on mental health (Calcia et al. 2016) and cognitive processes (Shields et al. 2017), decades of research has shown a plethora of negative side-effects of stress on the body, such as ulcers (Glavin et al. 1991), cardiovascular disorders (Dawson 2018), inflammation (Marsland et al. 2017) and even high incidences of suicidal thoughts (O'Connor et al. 2017) and self-harm. With estimates ranging as high as 74% of people reporting feeling stressed over the course of a year, there is an urgent need for better ways to assess and intervene against stress.

It is well known that stress resides in the brain, and is manifested throughout the body, influencing other biomarkers such as heart-rate variability (HRV) (Thayer et al. 2012). The aim of this study has been to go straight to the core to see if it would be possible to create a brain-based stress metric. By relying on the recent advances in technology and neuroscience we have tested the feasibility of measuring stress using neurophysiological measures. Moreover, in addition to using clinical-grade EEG systems (ABM X-10), a significant part of this study has been to test whether consumer electronics solutions such as the Muse headband and heart-rate variability could reliably track stress as it unfolds.

2 Methods

Participants were recruited through social media (e.g., Facebook) advertisements, and through Neurons' participant database. The test was performed in a central location (SingularityU Nordic, Titangade, Copenhagen, Denmark).

In total, 63 participants were recruited (age mean 30.9, st. dev = 5.5, 32 women). Each participant was equipped with both an Advanced Brain Monitoring (ABM) X-10 EEG brain monitor, and a Muse headband EEG headset. Via the ABM headset, we also recorded heart rate, which was then used to calculate HRV.

Participants received instructions about the study and the used methodologies, and signed an informed consent. All participant data were anonymized from the point of

© Springer Nature Switzerland AG 2020
D. D. Schmorrow and C. M. Fidopiastis (Eds.): HCII 2020, LNAI 12196, pp. 209–214, 2020.
https://doi.org/10.1007/978-3-030-50353-6_15

recording. All procedures were conducted in accordance with the local ethical committee (Videnskabsetisk Komite for Region Syddanmark) and followed the guidelines by the Helsinki Declaration.

Participants were positioned in front of a 27-in. computer monitor, and with a 14 in. second screen to their left side. The computer was running OpenSesame v3.2.7 in a Windows 10 environment. Most stimuli occurred on the screen in front of the participant, but during the stress test, a documentary was continuously shown on the secondary screen.

The trial was conducted in the same sequence for all participants:

- They were first given a set of different tasks such as watching a documentary, doing tasks related to arousal and working memory. This was not directly related to the study but helped to make the participants acquainted with the study setup and at the same time spend a few mental resources before the stress task started.
- During the stress phase of the trial, we used three load levels:

  - We had "Rest" phases between each task level.
  - In "Stress1" participants were asked to watch a documentary, and to pay attention to the content.
  - In "Stress2" they were asked to pay attention to the documentary while doing a memory task (They were presented to 3 digits and after 4 s they were asked to recite the digits in the same order).
  - In "Stress3" participants were asked to pay attention to the documentary while two memory tasks (They were presented to a pattern then 3 digits to remember, and after 4 s, they were asked to answer a question about the pattern and recite the digits in the same order).

- At the end of the trial, we included a final "Rest" phase.

After the trial, participants were unmounted with the equipment and given an online survey to complete. In this survey, they were asked about their experience of stress during each of the stress tasks, and asked to rate their level of experienced stress using a 7-point Likert scale ranging from 1 ("not stressed at all") to 7 ("extremely stressed").

For statistical analyses, we ran two main approaches. First, we tested previously reported findings as hypotheses. Second, we also ran multiple tests to test for optimal levels of prediction of stress levels. For each method, we ran a 50/50 split on training and testing. Respondent data were randomly assigned to one of two groups. The data from one group was used for training the model, and the second group was used to assess the test responses as the determining factor for evaluating the results of the model. The exploratory methods included Least absolute shrinkage and selection operator (LASSO), Least-angle regression (LARS), Partial Least Squares (PLS) regression, and deep neural network supervised machine learning analyses.

These analyses were run for both the ABM EEG and the Muse headband EEG. For the HRV measure we only ran hypothesis-specific analyses.

3   Results

In the analyses we made several significant findings. Here, we summarize the findings briefly:

- **Previous research is confirmed, with a limitation** – While we can confirm what previous studies have found about stress and the brain, these findings are not sufficiently reliable to be considered as sufficiently sensitive or reliable stress biometrics.
- **An 82% brain predictor of stress** – Using machine learning methods with the ABM EEG data, we have successfully found a model that can predict stress with 82% accuracy. This is close to clinical-grade accuracy of 90%.
- **A lightweight model is also predictive** – We also found that the Muse headband could reliably predict stress from brain signals with a 72% accuracy. Somewhat lower than the ABM EEG equipment. We can therefore successfully claim that a consumer electronics method can reliably predict stress. That said, since the data loss from the Muse was high (50%) it is recommended to identify a more reliable lightweight EEG device.
- **Heart rate measures preferred for lightweight measures** – Surprisingly, we find that heart rate variability (HRV) produced a strong predictive capacity to detect stress, with a 78% accuracy. This suggests that HRV alone, or in conjunction with the Muse headband, might work as a scalable stress toolkit.
- **Stress detection instead of stress levels** – One consistent finding across methods was that stress prediction of two levels (Stress vs Rest) produced the highest prediction scores compared to the original data with four levels of stress.

4   Limitations

It is worth mentioning that there was a substantial data loss rate from the lightweight EEG. On average, fifty percent of the recording time was lost due to poor acquisition from the Muse headband. A more reliable lightweight acquisition hardware needs to be identified to perform more robust measures.

One further limitation of the study regards the selection of participants. In this study, the data was acquired from a convenience sample of young adults, with ages ranging from 20 to 45 years old. Furthermore, conditions such as the previous stress level of the participants, chronic stress, or occurrence of stressful life events were not recorded from the participants, as a means to avoid unveiling the purpose of the research. Nonetheless, the study was designed to expose the participants to increasing stress levels, associating the stress response to the demands of the task, regardless of the previous mind state of the participant. Therefore, we are still able to link the neurometric and biometric markers to the level of stress associated with the task (Figs. 1, 2, 3 and 4).

Fig. 1. Equipment setup for stimuli exposure

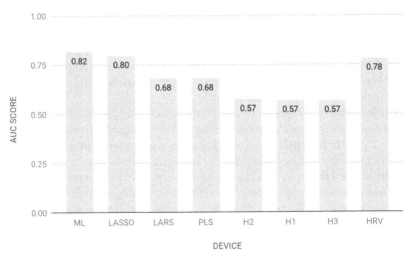

Fig. 2. Prediction power of each of the measures for the ABM X-10 device, and compared to the HRV measure (right). Bars denote the prediction accuracy of each score.

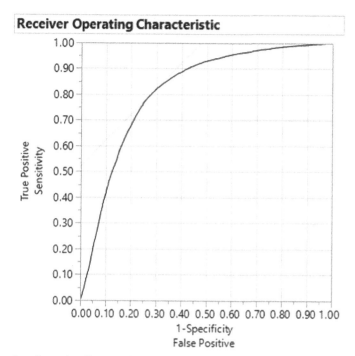

Fig. 3. Receiver Operating Curve (ROC) analysis for the best performing analysis: the machine learning model using the ABM X-10 EEG data, in which the Area Under the Curve (AUC) showed a value of 0.82.

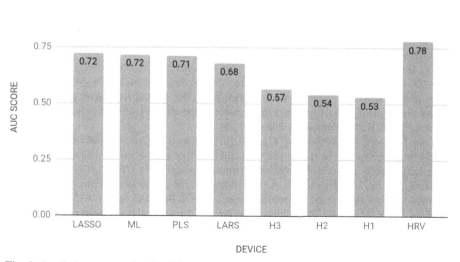

Fig. 4. Prediction power of each of the measures for the muse EEG, and compared to the HRV measure (right). Bars denote the prediction accuracy of each score.

5 Conclusion

The aim of the present study was twofold. First, we sought to identify the most reliable measures of stress. Here, we found that the ABM-based machine-learning model of stress could predict stress with an 82% accuracy, and the hypothesis-driven HRV measure was 78% accurate. This strongly implies that it is possible to use neurophysiological measures that are close to clinical accuracy (90%). The second aim of this study was to test the feasibility of using consumer-grade devices for tracking stress. As the HRV-based metric was highly sensitive and reliable, these results confirm the feasibility of such a solution. For the Muse-based EEG headset, we find that although the results are promising, there was far too much data loss to allow us to recommend this device as a recording device. However, recent data testing with the Muse-only now suggests that this approach is still feasible.

Taken together, this study demonstrates that portable stress metrics are possible. This fulfills the first criteria for successful stress intervention: detection. By having devices available for live and mobile stress measurements, it is now possible to speculate about possible uses of mobile stress metrics. For example, it might be possible as a solution for existing devices that meet certain minimum criteria, to be used in a closed-loop private system that empowers the individual. Second, it might be used in clinical settings for controlled intervention programs. Finally, such a solution can also be used (with the necessary ethical precautions) in work situations, in which larger companies can involve in "stress mapping" of anonymized individuals, and where interventions can be tested out. We believe that the present study lays the foundation for such next steps.

References

Calcia, M.A., Bonsall, D.R., Bloomfield, P.S., Selvaraj, S., Barichello, T., Howes, O.D.: Stress and neuroinflammation: a systematic review of the effects of stress on microglia and the implications for mental illness. Psychopharmacology **233**(9), 1637–1650 (2016). https://doi.org/10.1007/s00 213-016-4218-9

Dawson, D.: Acute stress-induced (takotsubo) cardiomyopathy. Heart **104**(2), 96–102 (2018)

Glavin, G., et al.: The neurobiology of stress ulcers. Brain Res. Rev. **16**(3), 301–343 (1991)

Marsland, A., Walsh, C., Lockwood, K., John-Henderson, N.: The effects of acute psychological stress on circulating and stimulated inflammatory markers: a systematic review and meta-analysis. Brain Behav. Immun. **64**, 208–219 (2017)

O'Connor, D., Green, J., Ferguson, E., O'Carroll, R., O'Connor, R.: Cortisol reactivity and suicidal behavior: investigating the role of hypothalamic-pituitary-adrenal axis responses to stress in suicide attempters and ideators. Psychoneuroendocrinology **75**, 183–191 (2017)

Shields, G., Sazma, M., McCullough, A., Yonelinas, A.: The effects of acute stress on episodic memory: a meta-analysis and integrative review. Psychol. Bull. **143**(6), 636–675 (2017)

Thayer, J., Åhs, F., Fredrikson, M., Sollers, J., Wager, T.: A meta-analysis of heart rate variability and neuroimaging studies: implications for heart rate variability as a marker of stress and health. Neurosci. Biobehav. Rev. **36**(2), 747–756 (2012)

AI and Augmented Cognition

No Free Lunch: Free at Last!

Ali Almashhadani[1], Neelang Parghi[2], Weihao Bi[2], and Raman Kannan[2(✉)]

[1] Hunter College, City University of New York, New York, USA
[2] CSE, Tandon School of Engineering, New York University, Brooklyn, NY 10201, USA
rk1750@nyu.edu

Abstract. No Free Lunch (NFL) Theorem in M/L is rigid and inflexible, which states that "No particular classifier can outperform all the other classifiers for every dataset". In this paper we present a MISD machine that runs multiple classifiers in parallel against a given dataset, implementing a "Swiss Army Knife" to combine many different classifiers and review their performance, effortlessly. The service will be hosted as a public service over the internet for any Machine Learning practitioner to experiment with datasets.

1 Supervised Learners

Machine Learning seeks to learn from known data and apply it to never seen before data. Classification or Supervised Learning is one of the core Machine Learning tasks. In Supervised Learning, one learns to assign a label (class) given a vector of predictors. Interested readers may find summary introduction in [1] and deep introduction in [2] and there are several other classics on the subject matter [3–5], and for those interested in managing large scale machine intelligence projects [6] is an excellent source (Fig. 1).

There are many Classification algorithms as shown above and one is faced with a dilemma: Which algorithm should one use given a particular dataset?

1.1 Satisficing Solution[x]

The satisficing decision-making as discussed in [6] is a heuristic where people settle with a solution to a problem that is 'good enough' but may not be the optimal one. A "Satisficing Solution" can be considered as a vernacular description of Occam's Razor [7, 8]. The notion of Satisficing Solution does not run counter to the well known axiom "No Free Lunch Theorem" [10] in Machine Learning. In combination with the razor, a satisficing solution is good enough.

1.2 No Free Lunch (NFL) Theorem

There is considerable debate [9, 10] about NFL [11] as to its meaning and interpretation and there is even an organization dedicated to NFL [12].

© Springer Nature Switzerland AG 2020
D. D. Schmorrow and C. M. Fidopiastis (Eds.): HCII 2020, LNAI 12196, pp. 217–225, 2020.
https://doi.org/10.1007/978-3-030-50353-6_16

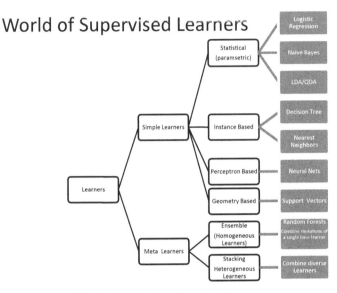

Fig. 1. Machine learning classifier tree

1.3   What Is Cost?

If it cannot be free, what is the cost? As outlined in [6] and [13], misclassification error is not the only cost. There are other costs including:

a) demand on memory,
b) processing time and
c) interpretability.

1.4   Need for Automated Algorithm Selector

In our opinion, as M/L is adopted more and more, the most impactful consideration for practitioners is that there is no single classifier can outperform in all domains. Consequently and it is imperative for practitions to ask the fundamental question posed in [14] "Among all the available classification algorithms, and in considering a specific type of data and cost, which is the best algorithm for my problem?" before settling on a particular algorithm. As the number of practitioners increase, ability to run a model will cease to be an advantage. The need for automating the algorithm selection process will become all too important and immediate. There have been several experiments comparing classifier performance [15–18], but none is available as a service to practitioners.

In this paper we will present our efforts, the Swiss Army Knife for No Free Lunch (NFL-SAK) to make lunch free for anyone with a dataset. Consistent with Occam's Razor, we allow users to submit a dataset, provide some hints to the structure of data and run several established classification algorithms of different types (parametric, instance based, logic based, ensemble and stacked-generalization). The NFL-SAK presents a useful tabulation of performance metrics. In its current form we present Area Under the Curve (AUC) [20] and Accuracy. There are several other performance metrics, see chapter 7 in Practical Data Mining [6] for a detailed overview and we plan to incorporate them in later revisions.

2 Implementation

Given a dataset, a model Formula, and a set of algorithms, NFL-SAK platform, performs a classification over the given set of algorithms. System uses readily available packages in R [21] including:

1. library (DMwR)
2. library (caret)
3. library (e1071)
4. library (pROC)
5. library (randomForest)
6. library (rattle)
7. library (rpart)

The process is intuitive as shown below (Fig. 2):

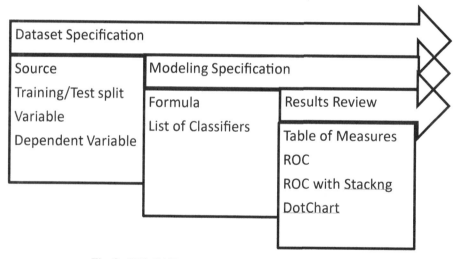

Fig. 2. NFL-SAK process and user interaction frameworks

The Shiny UI implements a "Classify By Example" model where the practitioner can specify One or more classifiers, the independent variable and the dependent variables. Consistent with Ockham's Razor, each selected classifier will be run with the simplest default model without parameter tuning and the results displayed for review (Fig. 3).

2.1 Dataset Specification

Here we have loaded the Hepatitis [29–31] dataset. We want to use 70% for training and the rest for testing.

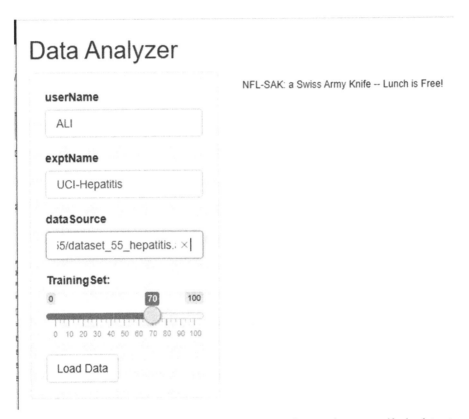

Fig. 3. Users first interaction with NFL-SAK, users name the experiment, specify the dataset.

2.2 Modeling Specification

Users can specify the model Formula and train one or more of the learners. Here we have identified the class variable and the list of learners we want to evaluate. Note that one parametric classification (Logistic Regression), one instance based classifier (kNN), one logic based classifier (Decision Tree) and a Support Vector Machine alongside RandomForest (an Ensemble classifier is given. Stacking with voting is also run by default) (Fig. 4).

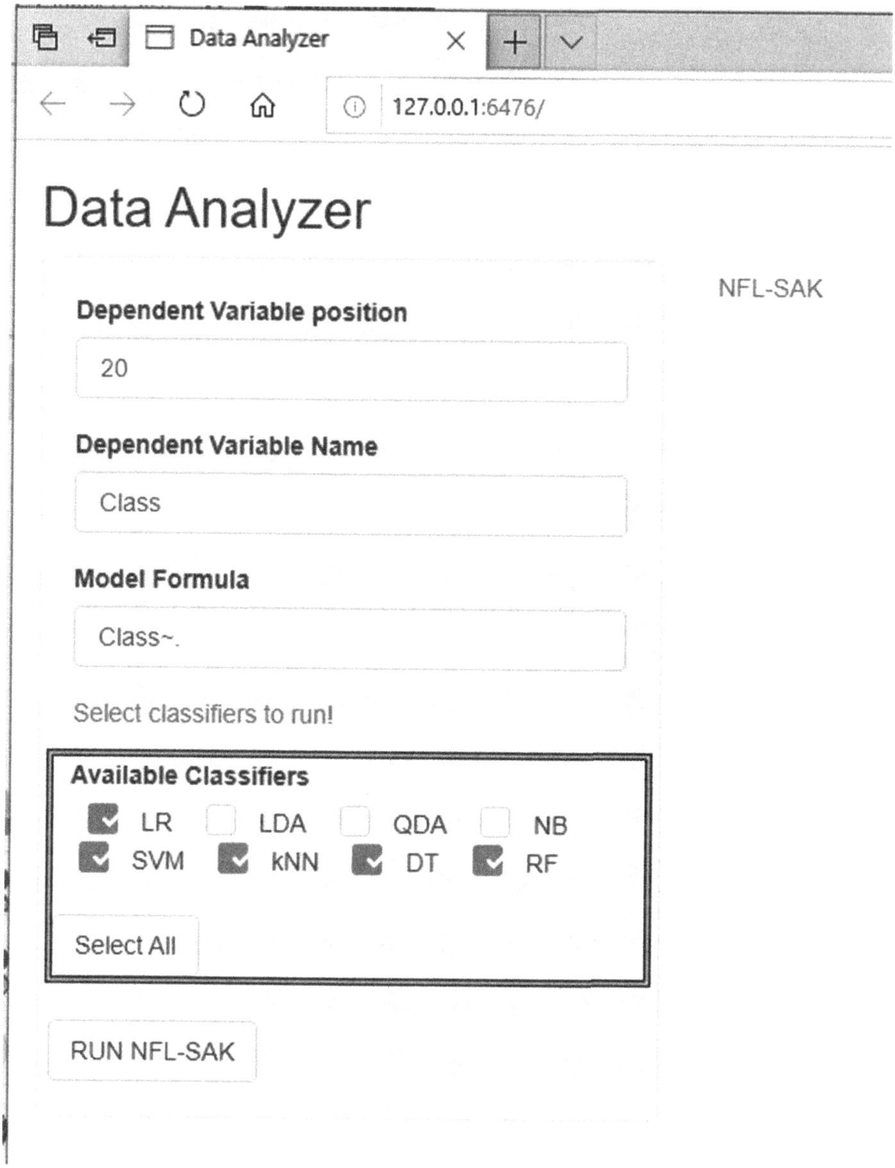

Fig. 4. Experimental design specification

Now we will run the model and review the results.

2.3 Model Output

First numeric performance measures including Accuracies and AUC are presented (Figs. 5 and 6).

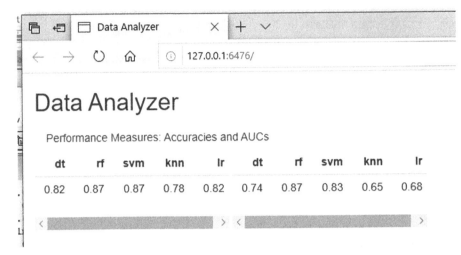

Fig. 5. Table of numeric performance metrics.

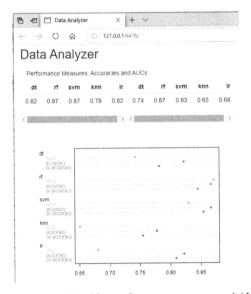

Fig. 6. Table of classifier performance accuracy and AUC.

Modest visualizations are presented allowing one to compare the relative measures. We used shiny [22] and shinyWidgets [23] for generating these visualizations and without the swarm wisdom available from netizens [32] none of this is possible, given that we are unfunded, staffed by 1 TA,1 Volunteer and 1 undergraduate student.

Results of stacked generalization is presented below (Fig. 7).

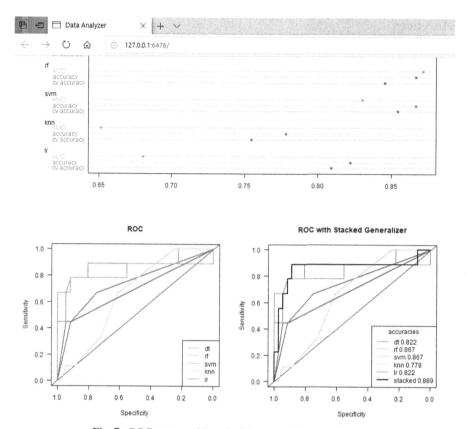

Fig. 7. ROC curves with and without stacked generalization

For the Hepatitis dataset, the stacked-generalizer using LogisticRegression, DecisionTree, Nearest Neighbor, Support Vector Machines, randomForest and the Random-Forest are shown as specified. The Stacked Generalizer results in the highest performance of 0.899 combining all the above classifiers including the Ensemble classifier.

3 Conclusion

In this paper we summarized the import of No Free Lunch theorem, efficacy of Occam's Razor in searching for the best performing classifier for any given dataset. Guided by Occam's Razor, weak learners are trained at default configuration. User is allowed to pick and choose algorithms, specify a training set proportion. The system then runs the stacked-generalizer using voting mechanism. Comparative performance measures are displayed with Accuracy and AUC. ROC curves are generated for the specified algorithms. Users can perform multiple experiments and save them for further analysis.

Acknowledgements. Generous support from IBM PowerSystems Academic Initiatives for all of Raman's course is acknowledged.

References

1. Kotsiantis, S.B.: Supervised machine learning: a review of classification techniques. Informatica **31**, 249–268 (2007)
2. Alpaydin, E.: Introduction to Machine Learning. MIT Press
3. Duda, R.O., Stork, D.G., Hart, P.E.: Pattern Classification. Wiley
4. Mitchell, T.: Machine learning. McGraw Hill
5. Murphy, K.P.: Machine Learning, A Probabilistic Perspective. MIT Press
6. Hancock Jr., M.F.: Practical Data Mining. CRC Press
7. Saisficing Solution. https://www.kbmanage.com/concept/satisficing
8. Domingo, P.: The role of occam's razor in knowledge discovery. Data Min. Knowl. Discov. **3**, 409–425 (1999)
9. No Free Lunch Theorem. https://medium.com/@LeonFedden/the-no-free-lunch-theorem-62ae2c3ed10c
10. https://peekaboo-vision.blogspot.com/2019/07/dont-cite-no-free-lunch-theorem.html
11. Wolpert, D.: The lack of a priori distinctions between learning algorithms. Neural Comput. **8**(7), 1341–1390 (1996)
12. http://no-free-lunch.org/
13. Turney, P.: Types of Cost in Inductive Concept Learning. https://arxiv.org/ftp/cs/papers/0212/0212034.pdf
14. Shilbayeh, S.A.: Cost Sensitive meta-learning. http://usir.salford.ac.uk/id/eprint/36278/1/Cost%20sensitive%20meta%20learning_2015.pdf
15. Salzberg, S.L.: On comparing classifiers: pitfalls to avoid and a recommended approach. Data Min. Knowl. Disc. **1**, 317–328 (1997). http://people.sabanciuniv.edu/~berrin/cs512/reading/salzberg-comparing-pitfalls.pdf
16. Dietterich, T.G.: Approximate statistical tests for comparing supervised classification learning algorithms. http://web.cs.iastate.edu/~honavar/dietterich98approximate.pdf
17. Demsar, J.: Statistical comparisons of classifiers over multiple data sets. J. Mach. Learn. Res. **7**, 1–30 (2006), http://jmlr.org/papers/volume7/demsar06a/demsar06a.pdf
18. Caruana, R., et al.: An Empirical Evaluation of Supervised Learning in HighDimensions. http://lowrank.net/nikos/pubs/empirical.pdf
19. https://www.wired.co.uk/article/master-algorithm-pedro-domingos
20. Fawcett, T.: An introduction to ROC analysis. Pattern Recogn. Lett. **27**, 861–874 (2006)
21. R Core Team: R: A language and environment for statistical computing. R Foundation for Statistical Computing, Vienna, Austria (2018). https://www.R-project.org/
22. Chang, W., Cheng, J., Allaire, J.J., Xie, Y., McPherson, J.: Shiny: Web Application Framework for R. R package version 1.2.0 (2018). https://CRAN.R-project.org/package=shiny
23. Perrier, V., Meyer, F., Granjon, D.: shinyWidgets: Custom Inputs Widgets for Shiny. R package version 0.5.0 (2019). https://CRAN.R-project.org/package=shinyWidgets
24. Robin, X., et al.: pROC: an open-source package for R and S + to analyze and compare ROC curves. BMC Bioinform. **12**, 77 (2011). https://doi.org/10.1186/1471-2105-12-77 http://www.biomedcentral.com/1471-2105/12/77/
25. Meyer, D., Dimitriadou, E., Hornik, K., Weingessel, A., Leisch, F. (2019). e1071: Misc Functions of the Department of Statistics, Probability Theory Group (Formerly: E1071), TU Wien. R package version 1.7-0.1. https://CRAN.R-project.org/package=e1071
26. Kuhn, M., et al.: caret: Classification and Regression Training. R package version 6.0-84 (2019). https://CRAN.R-project.org/package=caret
27. Therneau, T., Atkinson, B.: rpart: recursive Partitioning and Regression Trees. R package version 4.1-13 (2018). https://CRAN.R-project.org/package=rpart

28. Liaw, A., Wiener, M.: Classification and regression by randomForest. R News **2**(3), 18–22 (2002)
29. https://www.openml.org/data/get_csv/55/dataset_55_hepatitis.arff
30. https://archive.ics.uci.edu/ml/datasets/Hepatitis
31. https://github.com/datasets/hepatitis
32. https://stackoverflow.com/questions/34384907/how-can-put-multiple-plots-side-by-side-in-shiny-r

Biomimetic Design in Augmented Cognition

Benjamin Bowles[1]([✉]), Monte Hancock[2], Mitchell Kirshner[3], and Taniya Shaji[1]

[1] Sirius20, Melbourne, FL, USA
benbowles2@gmail.com
[2] 4Digital, Los Angeles, CA, USA
[3] University of Arizona, Tucson, AZ, USA

Abstract. Biomimetic ("life mimicking") systems automate functionality by replicating biological forms and processes (e.g., early airplane designs were trying to model winged flight in birds). In this way, organic systems not only supply proof of concept ("heavier-than-air flight is possible"), but also inform attempts to implement functionality by reproducing form in automated systems.

Important sub-disciplines within the science of machine learning have developed in this same way. The most widely used cognitive architectures closely resemble their biological starting points. Obvious examples are neural networks, expert systems, and genetic algorithms. The application of sophisticated optimization techniques to these systems can supersede unscalable aspects of the underlying biological metaphor ("airplanes do not flap their wings"). However, biomimetic principles inform possible approaches to solving many problems.

This paper describes the application of biomimetic design to augmented cognition and analyzes machine learning developments from a biomimetic lens. An experiment follows comparing Neural Networks and Bias-Based Learning and assesses their performance against human subjects. The last part synthesizes the results of the experiment with the principles underlying biomimetic design to foster effective problem-solving.

Keywords: Biomimetic · ML design · Machine learning

1 Biomimetic Systems

In machine learning, biomimetics provide a natural starting point due to the sophistication of the human brain and its ability to solve almost any problem. Inspiration from the brain can inform artificial intelligence (AI) by demonstrating successful cognitive tactics. In fact, all AI is biomimetic to some degree since intelligence is a biological concept.

However, biomimetics have not proved a panacea for every machine learning problem. Biomimetic inventions often suffer from obstacles in translating complex biological processes to a technical context. Addressing these roadblocks typically requires analysis of the problem and a non-biomimetic, experimental approach driven by nonbiological principles. Math has often proved essential in bridging the gap between these concepts and execution. While biomimetics are useful for understanding the problem and potential solutions, excessive commitment to the organic inspiration hinders augmented cognition.

© Springer Nature Switzerland AG 2020
D. D. Schmorrow and C. M. Fidopiastis (Eds.): HCII 2020, LNAI 12196, pp. 226–240, 2020.
https://doi.org/10.1007/978-3-030-50353-6_17

In flight, early inventors looked to the skies, where they found feathered existence proofs of extended flight. Many designers attempted to create what amounted to large bird exoskeletons for humans to flap. Only in hindsight is it clear that human-powered flight is a true challenge and that other methods are far more effective. Instead, this principle indicates a different goal for biomimetics: understanding the problem space to allow a creative solution.

1.1 Neural Networks

Neural networks constitute a cognitive model wherein many small "neurons" are connected to each other in several layers. In the most prolific, the feed-forward neural network, charges accrue from the first layer to the last layer, passing through many neurons and being modified by each. Neurons are connected to all neurons in the next layer, with each connection in the network given "weight," or specific parameter, by which the charge is multiplied, representing its importance [1] (see Fig. 1). When enough charge generates past the firing threshold, then the neurons fire passing the signal to the next layer. Each neuron has a certain "firing threshold" which affects the base likelihood of firing. Through this structure, simple neurons are capable of approximating complex solutions.

McCulloch and Pitts, a duo of scientists seeking to understand the process of thinking by modelling the form of the human brain, first popularized neural networks. They sought to imitate the structure of the brain for the purpose of psychological study, and their neural network was thus describing cognition rather than solving problems as they do today. This work contributed a meaningful biomimetic concept to the field of neural networks: function imitates form. As McCulloch writes, "There is no theory which will retain... if the net be altered" [2]. Changes in structure lead to differences in outcome.

Numerous other contributors slowly transformed neural networks from a model to a tool. Although the concept of problem-solving with neural networks arose quickly, the problem of training the many parameters of networks went unsolved for more than a decade. In the 1960s, however, rocket scientists Kelley and Bryson independently developed their methods for training neural weights. They "climbed the hill" to maximize network performance by taking small steps in the right direction in the n-dimensional topography of machine parameters [3]. However, the work of Kelley Bryson remains largely unrecognized.

In 1974, Paul Werbos advanced his Back Propagation algorithm, a gradient-based method for optimizing the parameters of neural networks [3]. In 1986, Rumelhart and Hinton described the process of Back Propagation in detail [4]. Although they indicated the method was not developed sufficiently for routine use, this solution became popular and allowed widespread usage of neural networks. Critically, the problem of training neural networks was solved by looking away from the brain and towards calculus. While Back Propagation was not, in Rumelhart's words, a "plausible model of learning in brains," it revolutionized artificial intelligence. However, some problems in neural networks are solved in ways reminiscent of their biological inspiration.

The "black box" phenomena occurs when machine decisions seem like a black box – the input and output are visible, but the connections between are unknown. One attempt to solve this problem is the performance manifold, an n-dimensional map of the parameter space making up the brain of a neural network. Recent experiments indicate slices of this space can be colorized by performance to indicate the potential pathologies in the network structure or data [5]. Performance manifolds are conceptually like MRIs of human brains and serve a similar purpose but are not designed to imitate the biological function. The underlying principles of performance manifolds were developed mathematically rather than biomimetically.

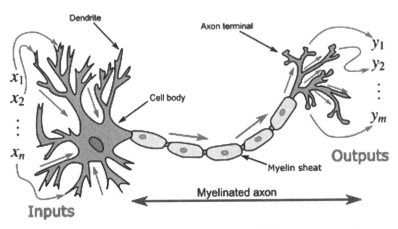

Fig. 1. Artificial neurons mimic the structure of biological neurons [6].

1.2 Knowledge Based Expert Systems

Knowledge-Based Expert Systems (KBES) are constructs for augmented cognition that can aid decision making in problems through rule-based reasoning, which represent knowledge from experts. Two basic forms of KBES exist: backward chaining (logic programming systems) and forward training (production systems). Backward chaining deduces the necessary past states based on the endpoint, and is often used for diagnosis. Forward chaining uses an opposite form of reasoning with production rules, which are triggered when its condition is met in a model of the problem (working storage). Forward chaining earns its name since the action of each rule applied by a rules engine updates the working storage model and affects the selection of the next rule [7].

The knowledge base and the inference engine compose the two main parts of the KBES. The knowledge base contains factual and heuristic knowledge, and the inference engine uses that knowledge base to construct the line of reasoning leading to the problem solution. As a form of artificial intelligence, KBES can be customized to different domain spaces through knowledge engineering. Commercial software exists to facilitate knowledge engineering by obtaining knowledge representation without explicitly programming rules. Current software fails to handle knowledge acquisition, which can be automated through machine learning for information extraction from experts, textbooks, and real-world data [8].

1.3   Bayesian Classifiers

Bayesian classifiers are artificial intelligence implementations, which rely on probability to estimate the likelihood of events. This method originated from the work of Thomas Bayes, a reverend who developed several theorems in an influential 1764 paper. Bayes' Theorem states that the probability of some event H given some event X is equal to the probability of X given H times the probability of H, divided by the probability of X (1) [9].

$$P(H|X) = \frac{P(X|H) * P(H)}{P(X)} \tag{1}$$

For example, the chance that a person is a farmer, given that they are male, is equal to the probability that a farmer is male times the probability of being a farmer, divided by the probability of being male.

Bayes' theorem launched an entire branch of statistics established on estimating probabilities based on evidence. Bayesian techniques accrue evidence, which increase belief in certain conclusions. This rational and evidence-based method is easily applied to AI. Given some number of classes, a Bayesian classifier will give the answer to a certain example it believes has the highest probability based on previous evidence [10].

Bayesian classifiers also make a flawed assumption to increase their practicality. Since the probability of a specific feature vector given a class is difficult to calculate, Naive Bayesian Classifiers assume independence of features. This allows them to estimate the probability an example is from a certain class much more rapidly.

From a biomimetic perspective, Bayesian classifiers are notable, because they attempt to model the function of human cognition. Bayes understood that 1) human cognition is an efficient means for estimating probability based on small samples, and 2) human thinking is based on iteratively changing beliefs based on all the available evidence.

In this way, Bayesian classifiers represent the antithesis of neural networks. Neural networks are based on a correct assumption (function follows form) which is difficult to implement and slow to approach an exact answer. Bayesian classifiers are based on a fundamentally flawed assumption of independence used to allow a rapid estimation of probability. Neither is inherently superior, but both methods implement unique solutions inspired by biological inspirations.

1.4   Bias-Based Reasoning

Bias-Based Reasoning (BBKBES) is a form of artificial intelligence, which uses trainable, automated Expert-Based Systems to rapidly memorize problem data [11]. Bias-based reasoning models the function of human cognition by accruing bias for and against conclusions based on experience, like Bayesian classifiers. However, BBKBES makes each rule trainable through "weights" which are treated as the reliability of that rule. In this way, Bias-Based Reasoning assigns nonbinary truth values (see Fuzzy Logic).

Because it lacks vast networks in a complex parameter space, BBKBES is faster than neural networks. This simplicity of structure also renders explainable decisions, in contrast to the "black box" phenomena of neural networks. Rules, which are natural to human thought, serve as clear justifications for conclusions.

BBKBES, which is currently in development, is an example of biomimetic design succeeding despite departing from its inspiration. Although the concept of BBKBES is taken from human thought, the actual mechanics, involving a grid in space and an algorithm to train the weights, are entirely original. This creativity borrowed an essential strength of the brain (accruing biases toward conclusions) and implemented it in a unique way accentuating the strengths of AI. Although this uniqueness is less biomimetic, it allows purposeful optimization of performance.

2 Biomimetic Techniques

2.1 Genetic Algorithms

Machine learning scientists employ genetic algorithms to optimize parameters of artificial intelligence [12]. Genetic algorithms implement some approximation of natural selection to ensure the most successful individuals become more prolific. The largest requirement for a genetic method is a genetic code. A standardized numerical encoding is required to allow genetic evolution to proceed. This is convenient for AI, where programs are typically represented by a complex set of parameters. For example, neural networks are parameterized by the weights connecting neurons.

A genetic optimization begins with a population where all parameters are instantiated randomly [12]. Next is the competition stage, wherein the individuals compete to maximize some objective function, like survival or regression of data. AI will typically maximize classification accuracy for a training dataset, although it may also perform a regression. Most of this random population will perform poorly on the dataset, but some will stand out and be favored in the reproduction stage.

Following the competition stage is reproduction, where a genetic algorithm is employed to select which individuals reproduce [12]. In general, the most successful individuals from competition will receive a significant bias toward reproduction. Like biological reproduction, parameters are most often passed on directly to the next generation. However, mutation occasionally modifies parameters in a random fashion. This combination of survival of the fittest with mutations allows genetic algorithms to evolve towards an optimal individual (see Fig. 2).

In effect, genetic algorithms traverse the parameter space of artificial intelligence in a manner like backpropagation of neural networks or other optimization methods. However, genetics employs randomness rather than the more systematic approach of gradient-based processes. This may allow genetic algorithms to avoid becoming trapped in local minima like hill-climbing optimizations. From a biomimetic perspective, genetic algorithms replicate a function so essential it is almost universal. Reproduction and evolution over many generations apply to both organic and artificial contexts because of selection's ability to quickly approach an optimal individual.

Repeat

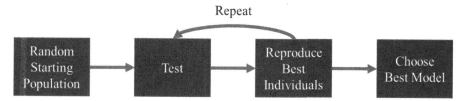

Fig. 2. The stages of genetic algorithms for machine learning

2.2 Problem-Solving Techniques

Artificial intelligences are often asked to solve specific problems. A successful artificial intelligence is expected to act dynamically, responding to the situation at hand. This unlocks a realm of possibilities, which can inform and revolutionize computing. Strategies for "teaching" AI to solve problems mimic techniques employed by humans.

Early in the history of AI, mazes and tree games were used to evolve effective algorithms [13]. This was based off the contemporary belief that decision trees modelled human decision making. Mazes were so popular that the term "problem maze" preceded our current concept of "problem space." However, limiting the proliferation of options under consideration is a major problem with large decision trees. Early techniques, which focused on human heuristics, declined in popularity as algorithms increased. Algorithms follow a ruleset guaranteed to eventually solve the problem.

Depth-first is a simple algorithm, which goes as far as possible in a direction and retraces its steps in the maze when blocked [14]. However, depth-first strategy is inefficient and does not fully explore the problem space. An alternative to depth-first is breadth-first, which systematically explores the entire space and determines the optimal path. Breadth-first is more efficient than depth-first due to avoiding backtracking and guarantees the ideal solution. One evolution is iterative depth-first, which runs multiple times with increasing depth limits, allowing for greater efficiency.

A more complex method is A*, a combination of algorithm and heuristic which favors paths closer to the end goal of the maze [14]. The specific function favoring these closer areas may be adjusted depending on the specifics of the problem. A* combines the systematic capability of algorithms to guarantee a solution with the efficiency heuristics offer (See Table 1). Beyond mazes, other AI strategies imitate human approaches.

Pruning is a strategy suited for games represented by decision trees. In pruning, a program simulates the logic of a game in order to evaluate specific strategies and rule out methods that should not be used. For instance, in the game of Nim, each player takes either one or two sticks and attempts to avoid taking the last stick. To prune Nim, the program steps backwards from one stick left (which is a fatal position) and evaluates the capability of each position to force the player into a loss. Two and three sticks are optimal positions since they allow the player to leave their opponent with one stick, winning the game. Pruning generally assumes optimal counter play, which is why four sticks cause a fatal position [15]. They lead to either two or three sticks, allowing the opponent to force the player's loss. This method can increase efficiency and avoid the worst pitfalls available.

All these problem-solving techniques strive to decrease the number of options under consideration. Ruling out choices is a fundamental aspect of human decision-making. However, these algorithms departed from human heuristics in order to develop reliable and efficient methods of searching. Human processes inspired algorithms, but biomimetics alone did not offer a solution to decision tree problems.

Table 1. Performance of various tree search algorithms [14].

Search algorithm	Performance	Complete	Optimal
Breadth-first	High	Yes	Yes
Depth-first	Low	No	No
Iterative depth-first	Medium	Yes	Yes
(A Star)	Very high	Yes	Yes

2.3 Swarm Intelligence

Swarm intelligence is a metaheuristic approach to solving combinatorial optimization problems. Swarm intelligence uses population-based stochastic methods based on the collective behavior of simple individuals, stemming from interactions with the environment to produce population patterns. Many swarm intelligence approaches are biomimetic. One example is ant colony optimization, where artificial ants travel through a problem graph to deposit digital pheromones which enable other ants to solve more optimal problems. This technique has solved the traveling salesman problem as well as network optimizations [16]. Another example is the Bees algorithm, inspired by honeybees, which models foraging behavior to optimize functions [17].

In addition to successful optimization, swarm intelligence holds promise for development of control schemes in a variety of computer network applications. Communication networks could make use of swarm intelligence to make better routing decisions based on emergent properties observed in the data [18]. Other multi-agent systems such as botnets have also proven compatible with swarm intelligence algorithms for control. Botnets can be designed for spontaneous, implicit collaboration between independent bot agents instead of relying on a master-slave architecture. Swarm intelligent botnets are robust, scalable, and more dynamic than traditional systems. They accomplish this by propagating control messages to bot nodes using short-range hops defined by a biomimetic calculation inspired by ant foraging [19].

2.4 Fuzzy Logic

Fuzzy logic is a method of representing statements as neither entirely true nor entirely false [20] (see Fig. 3). Unlike boolean crisp logic, fuzzy logic qualifies every proposition with a truth value between 0 and 1, non-inclusive. For example, the statement "Dogs have exactly three legs," while far from true, is not entirely false. Some dogs are three-legged. In fuzzy logic, this statement would be modified by a term like "Some." Although conceptually more laborious, fuzzy logic is ideal for deductive reasoning.

Fuzzy logic is based on a realization that the human practice of representing truth with qualifiers is useful for creating rules. Since absolute truth is not required, rules may be written which are only likely to be true, and the belief in those rules may be trained in a Bayesian fashion. This is perfectly suited to Expert-Based Systems (see Bias-Based Learning). Fuzzy logic is also easily explained through language. For instance, a fuzzy belief function can be converted into human inference. For this reason, Lotfi Zadeh, a pioneer in the field of fuzzy logic, termed it "Computing with Words" [21].

From a biomimetic perspective, the areas in which fuzzy logic both adheres to and differs from human thinking are purposeful. The qualifier-based system, which is directly inspired by the brain, is supported by numbers and equations, which are not evident in human thinking. To implement this usable inference framework, fuzzy logic largely does not use words as a basis for inference. Although words connect to fuzzy logic, those are explanatory conversions rather than the actual structure of the logic. This makes fuzzy logic far easier to implement and more flexible since numbers are universally applicable, although it differs from the apparent language of human cognition. Fuzzy logic is used as a tool to implement many machine learning methods, as a mathematical means of explaining conclusions and representing reliability of rules.

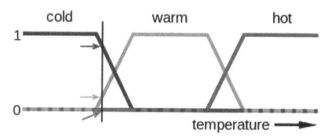

Fig. 3. In fuzzy logic, truth exists on a continuous scale rather than Boolean values [22].

3   Biomimetic Problems

3.1   Viruses and Worms

Malware refers to a wide array of programs designed to infect the host computers in which they live. The most famous malware, and the most biomimetic, are viruses, computer programs that live off another program and are narrowly tailored to carry out a specific purpose. Like biological viruses, computer viruses are often simple and, to succeed, they must spread.

Viruses will spread once executed either throughout the infected system or to other systems [23]. Like their organic counterparts, computer viruses hijack the machinery of their hosts, taking advantage of the resources available. Viruses typically propagate by appending themselves to other programs, which are then dispersed on the internet. For instance, many viruses are macros, small programs attached to common Microsoft files [24]. Infection may also include installation to a flash drive, sending spam emails containing the virus, or many other schemes.

However, computer viruses diverge from biological by intentionally harming their host. Once they are installed and activated, viruses typically steal information or install further malware to inflict any number of maladies on the host system. To accomplish their purpose of stealing money or information for their creators, computer viruses do not adhere to their model ambivalent treatment of their host cell.

Another type of malware is the worm, which differs since it infiltrates on its own initiative [25]. The term worm originates from the 1975 science fiction book The Shockwave Rider, which featured a "tapeworm" which actively spread between computers [26]. Worms will "burrow into" the hosts, like a parasitic worm [27]. Worms look for security flaws, which allow them to infiltrate more easily, typically within a host computer's operating system or software. Because they spread actively, worms can gain control of many computers very quickly. In fact, successful worms often spread so rapidly they congest the Internet [28].

For example, the CodeRed worm peaked at gaining two thousand hosts per minute [29]. Worms take advantage of the strengths of their organic counterparts to be incredibly successful since infection on the internet is functionally like biological parasites, allowing them to cause massive damage (see Fig. 4). Even though they were popularized by a science fiction story, divorcing their origins from reality, worms employ an effective biomimetic model to maximize their potential spread rate.

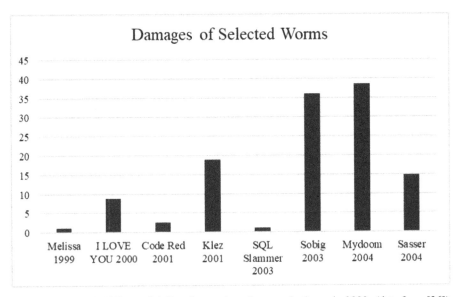

Fig. 4. Damages in billions of dollars from selected worms in the early 2000s (data from [26]).

3.2 Learning Pathologies

Overtraining is a machine learning problem where a model excellently analyzes training sets but fails to generalize this knowledge to the wider problem set. This may arise from a faulty training set, a problem common to human experience. On some occasions, the training data is insufficient and does not contain the information needed for a solution. Bad training data fails to represent the patterns, which solve the problem. Programs are still capable of memorizing the exact details of flawed training data. In the end, the model will accurately learn the random noise inside the data rather than the problem's solution.

Overtraining may also be caused by excessive time spent memorizing the data. Overtraining is essentially a problem unique to machine learning. This eventually trains the machine not to recognize problems *like* the sample data, but to memorize *exactly* the sample data. Due to the ability of computers to converge on specific values, too narrow modelling lessens the power of prediction. To avoid this, programs should stop at an arbitrary point before overtraining occurs, even if the errors on the training data are still decreasing (see Fig. 5) [30]. Unfortunately, this point varies per problem and is detectable primarily through evaluating a test set after the model is finished.

Even with models, which closely mirror human cognition like neural networks, overtraining remains a unique problem. Unfortunately, biomimetic methods cannot be applied to overtraining. Instead, overtraining is best addressed using effectual and representative training data, and through checking with a blind test set. This has become standard practice, marking another divergence from biological models needed to address context-specific problems.

Cognitive dissonance is the process by which a cognitive element becomes more confident in its conclusions, despite evidence remaining static [31]. Humans also fall victim to this process because of reducing uncertainty and embracing unwarranted conclusions.

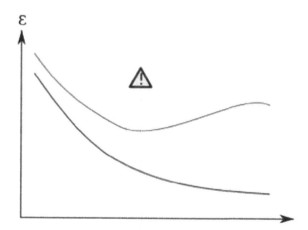

Fig. 5. Although error on the training set (blue) decreases continuously, error on the overall problem (red) eventually increases [32]. (Color figure online)

The structure through which humans evaluate their cognition values confidence. Unfortunately, this applies directly to many forms of machine learning, especially neural networks. Since the ideal network output is one hundred percent confident, neural networks will often boost their performance statistics through becoming incredibly confident.

This problem is an unfortunate side effect of the reinforcement method borrowed from human cognition. For instance, a program, which classifies the letters of the alphabet, will likely have similar confidence in an erroneous answer if a numerical digit, which it is not trained for, is provided. In cases like this, biomimetics can introduce negative aspects of their biological inspiration.

4   Experiment

4.1   Premise and Data

To compare the processing of biomimetic artificial intelligence and human cognition, an experiment was conducted wherein both human subjects and programs were asked to properly classify which side had the advantage in a tic-tac-toe game. This was a test of understanding of probability within the context of a simple game. For the classifiers, the data was fed in through a CSV file where one feature represented each board square (0 = Empty, 1 = X, 2 = O), and one feature was a flag indicating which side moved next (see Fig. 6). It is worth noting that every human subject came with tic-tac-toe experience, giving them an advantage over the classifiers in competition with them.

The ground truth for classification was determined by a stochastic simulation method wherein each board was played one million times with random choices. Then, the ground

Fig. 6. Side-by-side of machine data for training and human-readable test data. ID, turn, and class labelled for reader (original source is C45 numerical format).

truth was designated either 1 for X or 2 for O, corresponding to the side which won most. There were 1,512 boards, representing the 756 possible boards where each side had placed twice, expanded by a factor of two to allow a case where each side goes next.

4.2 Human Subjects

A survey of twenty-four high school seniors served as a control group in the tic-tac-toe experiment. The seniors were each presented with ten randomly chosen tic-tac-toe games where both sides had placed twice, as well as information designating the next turn. They were then asked to guess which side was more likely to win assuming they played randomly. The subjects answered seventy eight out of two hundred and forty questions correctly, averaging 67.5% accuracy.

4.3 Bias-Based Reasoning

A Bias-Based Reasoning classifier was applied to the same problem. The classifier reached 80.5% accuracy after five epochs, at which point it stagnated. This training process was completed within a minute. It is notable that this method significantly exceeded human classifications, despite lacking any understanding of the overarching rules of tic-tac-toe.

This success is also notable since one of the underlying assumptions of Bias-Based Reasoning was clearly violated. This problem of necessity did not maintain feature independence. Since the four placements are chosen from nine possible spaces, every problem exceeds the ten-percent guideline, violating independence. This may have still negatively influenced the classifier. For instance, X's in spaces one and nine are much more valuable together than separately, a fact this system may struggle to comprehend.

4.4 Neural Network

A neural network was also applied to the same problem. The network was composed of five hidden layers with five neurons, and an input layer of five neurons. After an initial 72% accuracy, the neural network converged to around 79.5% accuracy after several hours. While this is slightly less impressive than Bias-Based Reasoning, it still significantly beats the human sample.

5 Analysis

Both classifiers performed favorably on this data set. In fact, considering that randomness would yield approximately 50% accuracy, the programs both deduced approximately 60% of the difficulty of the problem, whereas the human sample solved only 30% of the problem's complexity. Considering the lack of background knowledge, this is particularly surprising.

This example shows the capability of biomimetic systems to surpass their inspirations. Since both the classifiers were generalized and lacked the mechanical game-knowledge of tic-tac-toe, their ability to synthesize the rules of the game indicates their ability to recognize the patterns in data. The fact both classifiers, using completely different approaches, converged on 80% accuracy indicates a tic-tac-toe game is only 80% solvable after four placements.

In this example, the programs were able to analyze the boards more fully because of their computing power. The subjects had a more complete dataset, considering extensive experience with the game. The humans possessed the theoretical capability to imagine every possible board, including those with more or fewer than four placements. However, they did not have the processing speed to derive the full context of the problem.

This concept that context can grant advantages or drawbacks to cognitive methods is crucial for understanding biomimetic design. Context shifts are critical and understood through retrospective analysis. Because of this unpredictable variable, unquestioning conformity to cognitive models may hinder or even destroy a design.

In cognition, applying effective biomimetics involves analyzing a specific function of human thinking and its strengths. Then, an artificial system is created to fill the same functions. This system does not need to copy human brains to benefit from the inspiration. For example, Bias-Based Reasoning is a much less direct approach to modeling human cognition than neural networks. This allows a much greater abstraction and some optimization, as evidenced by the significantly faster training time of Bias-Based systems.

Neural networks are robust systems able to approximate solutions to a vast set of problems efficiently. This problem-solving tool arose directly from a model of the brain. However, non-biomimetic strategies drive successful implementation. While neural networks have proven phenomenally successful, the divergences such as back-propagation have fueled this success.

In the case of flight, birds flap their wings to generate updrafts and sustain their flight. Birds also are light and streamlined to minimize drag. The first successful fliers, Wilbur and Orville Wright, took a more systematic and mathematical approach to flight. Based on their understanding of the forces involved, the Wright brothers calculated the necessary parameters for their flier [33].

Their largest abstraction, the use of an engine, was far more efficient than birds. From a removed point of view, the entire biological and caloric system of the bird serves as an engine for converting energy into power. Since it lacked a brain, the Wright brothers' twelve horsepower engine was a more specialized system.

This creative approach allowed the bypassing of critical inefficiencies, which had hamstrung flight for centuries. In this case, the outlined method allowed a superior understanding of the situation and led to a more optimized solution. As in flight, a rational approach to biomimetics requires avoiding the pitfalls inherent in basing a design on a biological foundation without considering mathematical alternatives. In cognition, biomimetics uses the inspiration to understand the problem, but a thoughtful designer must create the optimal solution.

References

1. Svozil, D., Kvasnicka, V., Pospichal, J.: Introduction to multi-layer feed-forward neural networks. Chemom. Intell. Lab. Syst. **39**(1), 43–62 (1997)

2. McCulloch, W.S., Pitts, W.: A logical calculus of the ideas immanent in nervous activity. Bull. Math. Biophys. **5**(4), 115–133 (1943)
3. Dreyfus, S.E.: Artificial neural networks, back propagation, and the Kelley-Bryson gradient procedure. J. Guid. Control Dyn. **13**(5), 926–928 (1990)
4. Rumelhart, D.E., Hinton, G.E., Williams, R.J.: Learning representations by back-propagating errors. Nature **323**(6088), 533–536 (1986)
5. Hancock, M., et al.: Visualizing parameter spaces of deep-learning machines. In: Schmorrow, D.D., Fidopiastis, C.M. (eds.) HCII 2019. LNCS (LNAI), vol. 11580, pp. 192–210. Springer, Cham (2019). https://doi.org/10.1007/978-3-030-22419-6_15
6. Vu-Quoc, L.: Neuron and myelinated axon, with signal flow from inputs at dendrites to outputs at axon terminals. Wikimedia Commons, Wikipedia. https://commons.wikimedia.org/wiki/File:Neuron3.png. Accessed 16 Dec 2018
7. Rules for Actions and Constraints - ScienceDirect. https://www.sciencedirect.com/science/article/pii/B9780128051603000053. Accessed 25 Jan 2020
8. CiteSeerX—Expert Systems: Principles and Practice. http://citeseerx.ist.psu.edu/viewdoc/summary??doi=10.1.1.34.9207. Accessed 25 Jan 2020
9. Joyce, J.: Bayes' Theorem. In: Zalta, E.N. (ed.) The Stanford Encyclopedia of Philosophy (Spring 2019 Edition). https://plato.stanford.edu/archives/spr2019/entries/bayes-theorem/
10. Leung, K.M.: Naive Bayesian classifier. Polytechnic University Department of Computer Science/Finance and Risk Engineering (2007)
11. Hancock, M.: Non-monotonic, bias-based reasoning under uncertainty. In: Proceedings of the 22nd International Conference on Human-Computer Interaction, Copenhagen, Denmark (July 2020)
12. Koza, J.R., Poli, R.: Genetic programming. In: Burke, E.K., Kendall, G. (eds.) Search Methodologies. Springer, Boston (2005). https://doi.org/10.1007/0-387-28356-0_5
13. Cordeschi, R.: Searching in a maze, in search of knowledge: issues in early artificial intelligence. In: Stock, O., Schaerf, M. (eds.) Reasoning, Action and Interaction in AI Theories and Systems. LNCS (LNAI), vol. 4155, pp. 1–23. Springer, Heidelberg (2006). https://doi.org/10.1007/11829263_1
14. Kaur, N.K.S.: A review of various maze solving algorithms based on graph theory. IJSRD, **6**(12), 431–434 (2019). ISSN (online): 2321-0613
15. Knuth, D.E., Moore, R.W.: An analysis of alpha-beta pruning. Artif. Intell. **6**(4), 293–326 (1975)
16. Swarms and Swarm Intelligence - IEEE Journals & Magazine. https://ieeexplore.ieee.org/abstract/document/4160239. Accessed 25 Jan 2020
17. Honey Bees Inspired Optimization Method: The Bees Algorithm. https://www.ncbi.nlm.nih.gov/pmc/articles/PMC4553508/. Accessed 25 Jan 2020
18. Swarm intelligence for routing in communication networks - IEEE Conference Publication. https://ieeexplore.ieee.org/abstract/document/966355. Accessed 25 Jan 2020
19. A botnet-based command and control approach relying on swarm intelligence - ScienceDirect. https://www.sciencedirect.com/science/article/pii/S1084804513001161. Accessed 25 Jan 2020
20. Zadeh, L.A.: Fuzzy logic. Computer **21**(4), 83–93 (1988)
21. Zadeh, L.A.: Fuzzy logic = computing with words. In: Zadeh, L.A., Kacprzyk, J. (eds.) Computing with Words in Information/Intelligent Systems 1. Studies in Fuzziness and Soft Computing, vol. 33. Physica, Heidelberg (1999). https://doi.org/10.1007/978-3-7908-1873-4_1
22. Shaw, R.S.: Wikimedia. Wikimedia, Wikipedia. https://commons.wikimedia.org/wiki/File:Warm_fuzzy_logic_member_function.gif. Accessed 15 Nov 2004
23. Subramanya, S.R., Lakshminarasimhan, N.: Computer viruses. IEEE Potentials **20**(4), 16–19 (2001). https://doi.org/10.1109/45.969588

24. Serazzi, G., Zanero, S.: Computer virus propagation models. In: Calzarossa, M.C., Gelenbe, E. (eds.) MASCOTS 2003. LNCS, vol. 2965, pp. 26–50. Springer, Heidelberg (2004). https://doi.org/10.1007/978-3-540-24663-3_2
25. Denning, P.J.: Computer viruses. Research Institute for Advanced Computer Science (1988). https://ntrs.nasa.gov/archive/nasa/casi.ntrs.nasa.gov/19890017050.pdf
26. Fosnock, C.: Computer Worms: Past, Present, and Future, vol. 8. East Carolina University, Greenville (2005)
27. Haas, W.: Parasitic worms: strategies of host finding, recognition and invasion. Zoology 106(4), 349–364 (2003). https://doi.org/10.1078/0944-2006-00125
28. Zou, C.C., Gong, W., Towsley, D.: Code red worm propagation modeling and analysis. In: Proceedings of the 9th ACM Conference on Computer and Communications Security. ACM (2002)
29. Moore, D., Shannon, C., Claffy, K.: Code-red: a case study on the spread and victims of an Internet worm. In: Proceedings of the 2nd ACM SIGCOMM Workshop on Internet measurement (2002)
30. Tetko, I.V., Livingstone, D.J., Luik, A.I.: Neural network studies. 1. Comparison of overfitting and overtraining. J. Chem. Inf. Comput. Sci. 35(5), 826–833 (1995)
31. Festinger, L.: Cognitive dissonance. Sci. Am. 207(4), 93–106 (1962)
32. Ratio Between Training Error and Validation Error. Stack Exchange. https://i.stack.imgur.com/HxlMa.png. Accessed 1 Aug 2015
33. Designing the Flier. https://airandspace.si.edu/exhibitions/wright-brothers/online/fly/1903/designing.cfm. Accessed 8 Jan 2020

A Field Theory for Multi-dimensional Scaling

Monte Hancock[1], Nick Nuon[2], Marie Tree[2], Benjamin Bowles[2(✉)], and Toni Hadgis[2]

[1] 4Digital Inc., Los Angeles, CA, USA
mlgodio@hotmail.com
[2] Sirius20, Melbourne, FL, USA
benbowles2@gmail.com

Abstract. An approach to multi-dimensional scaling is described which employs an analogy from the physics of conservative vector fields. This analogy allows the introduction of kinematic concepts into the data science problem in a natural way. Specific examples are presented.

The method described here uses multi-dimensional scaling to introduce information redundantly into feature sets for classifier problems. This is empirically shown to have beneficial effects for certain difficult classification problems.

This extends work done previously [1, 2] by using the posited physical analogy to make training more intuitive, efficient, and effective. A concept of super features is introduced and shown to improve classifier performance.

Keywords: Information theory · Torgerson Coordinates · Multi-dimensional scaling · Super features

1 Background

Classical Multi-Dimensional Scaling was introduced by Torgerson in 1952 [3] as a method of dimension reduction. In a vector space of dimension N, a distance matrix under some metric is constructed (row I, column J is the distance between points I and J in the data set.) The question is then asked, "Is there a set of vectors in a vector space of dimension M < N having this same distance matrix in some metric?" If so, this data set can serve as a lower-dimensional proxy which has the same metric structure (inter-vector distances, cluster structure, etc.) as the higher dimensional set.

The distance matrix serves as a dimensionless characterization of the data, and suggests many avenues for data science research. When such a proxy set can be created, coordinatizing it by specific registration with respect to some basis in the target space produces a set of "Torgerson Coordinates" for the new data.

Having to satisfy the Triangle Inequality means that it will not always be possible to find a lower-dimensional proxy set having exactly the same distance matrix as the original, higher-dimensional data set. In general, the greater the disparity between the original and target space dimensions, the greater will be the approximation error. (Of course, it is always possible to find perfect proxies in spaces having the original space as a subspace.)

D. D. Schmorrow and C. M. Fidopiastis (Eds.): HCII 2020, LNAI 12196, pp. 241–249, 2020.
https://doi.org/10.1007/978-3-030-50353-6_18

In practice, Torgerson Coordinates are generally determined using the Singular Value Decomposition. However, this approach does not provide the opportunities for intuitive experimentation provided by the field-theoretic approach, such as having a natural way to select different metrics in the source and destination spaces. Further, we have found that observing the data coalesce to the Lagrangian points under the action of the field derived from the distance matrix aids in developing intuition about the problem space.

- Physical kinematics models the interactions of material bodies that are immersed in fields which give rise to impressed forces. We employ an analogous
- Data Kinematics, models information binding across the components of multi-dimensional spaces. The theory is suggested by the following analogy:
- The Field-Theoretic Approach to the *Electrostatic and Gravitational Problem*: simplify and unify the complex problem of modeling many dynamically interacting things (e.g., free electrical charges moving in space-time), by modeling **one** thing: the field they generate.
- The Field-Theoretic Approach to the *Data Modeling Problem*: simplify and unify the complex problem of modeling many interrelated feature vectors (e.g., patterns emerging in feature data through a multi-dimensional information binding) by modeling **one** thing: a field they generate.

To mathematize the analogy, a data collection (e.g., a set of points in a Euclidean space) is considered to be Lagrangian points of a conservative vector field arising from discrepancies between the current proxy distribution and that expressed by the distance matrix of the original data. In keeping with the field-theoretic view, we accomplish this by numerically solving Laplace's Equation with boundary conditions. From a machine learning perspective, the boundary conditions expressed by the distance matrix provide training information for supervised learning. Properly defined, the resulting field model is conservative, satisfies the Superposition Principle, and is *approximately* unique up to rotation and translation.

2   The Distance Matrix

Let A be an ordered collection of vectors in $\mathbf{R^N}$:

$$A = \{\mathbf{V_1}, \mathbf{V_2}, \ldots, \mathbf{V_M}\}$$

where

$$V_j = \{V_{j1}, V_{j2}, \ldots, V_{jN}\}$$

In machine learning parlance, when the v_{ji} are attributes ("features") of an entity which V_j is intended to characterize, it is customary to refer to V_j as the entity's feature vector.

Let d be a rotation and translation-invariant metric ("distance function") on $\mathbf{R^N}$, and denote by d[A] the M-by-M matrix of pairwise distances given by d between the vectors of A. The row j, column i entry of d[A] is $d(\mathbf{V_j}, \mathbf{V_i})$.

It is customary to refer to d[A] as the distance matrix (sometimes the dissimilarity matrix) of A. The fact that A is ordered eliminates any possibility of ambiguity. d[A] is symmetric, non-negative, and has 0's on the diagonal.

All rotations and rigid translations of A will give rise to the same distance matrix. Further, d[A] encodes the fundamental structure of A in $\mathbf{R}^N$ in the sense that often the locations of a small number of the vectors of A determine the locations of all the vectors in A. (However, pathological cases for which this is not true can be created.)

More formally, we proceed as follows (Fig. 1):

Let $\mathcal{F} = \{A_j\} = \{A_1, A_2, \cdots, A_M\} \in \mathbf{R}^N$

Let $d_{ij}(A_i, A_j) = d_{ij}$ be a metric . Form the distance matrix:

$$D(A_i, A_j) = [d_{ij}], \qquad i, j = 1, 2, \ldots M$$

This matrix will be symmetric, zero diagonal, and non − negative.

Let: $\qquad S = \{\vec{a_j}\} = \{\vec{a_1}, \vec{a_2}, \cdots, \vec{a_M}\} \in \mathbf{R}^N$

be a (hypothetical) set of vectors having distance matrix D. Regarding the $\vec{a_j}$ as field sources, we define a discrete scalar potential $\wp$ on S by:

$$\wp(\vec{a_i}) = g \sum_{j=1}^{M} \left(\|\vec{a_i} - \vec{a_j}\| - d_{ij} \right)^2$$

Fig. 1. The field arises from discrepancies between a proxy set and original set

Two of the case studies cited above applied this theory to the analysis of social media spaces for application development. This work extends the application to modeling and classification of dynamic social media behaviors in context.

3   A Field-Theoretic Approach

The physical analogy enters our work in the following way. Suppose that the original N-dimensional data set consists of L vectors, and it is desired to find a corresponding set of L vectors in an M-dimensional space, where M < N.

The distance matrix provides a set of L − 1 requirements that each point in the lower-dimensional space must satisfy.

Consider proxy points I and J in the lower-dimensional space, initially placed randomly. Their actual distance from each other in the low-dimensional space might differ from the distance given in the original distance matrix Impose a force (proportional to the distance error) on these two points that "pushes them apart" if the actual distance is too small, and "pushes them together" if the actual distance is too great. Use the superposition principle to sum the forces on each point imposed by their discrepant distances from all other points. Since the force is scaled by the size of the discrepancy, this is an instance of the classical Delta Rule.

Each proxy point feels a force arising from its relative distance from all other proxy points. This can be thought of as a conservative vector field [1]. Allowing this force to act recursively through many epochs upon all the points will tend to arrange them so that the resulting distance matrix matches the original distance matrix. This corresponds to positioning the proxies at the Lagrangian points of this field, where all forces are zero (because all distance discrepancies are zero).

This optimization process can be used to carry out the recasting by fitting a set of exemplars to a set of reference vectors for the original feature vectors. The field-theoretic approach to determining the Torgerson coordinates is straightforward. Many implementations are possible. We originally proceeded as follows:

- For each vector in the original space, create a candidate proxy in the target space.
- To compute a correction delta for a vector, refer to the desired distance matrix. The candidate vector is too close to the candidates for others vectors, and too far from the candidates for others. Create a vector by superposing the errors in distance for all the other candidates. Add a scaled version of this "correction delta" to the vector.
- Do this for all vectors; this constitutes an update epoch.
- In actual practice, this approach is problematic. Vectors can be "blocked" by intervening candidates they need to avoid, and the computational complexity is quadratic in the number of vectors.

Using a low cardinality frame of vectors (e.g., the standard basis vectors) as a reference sets in the original data solved these problems. Update deltas need only be computed with respect to the reference vectors. This reduces the computation to linear complexity, and virtually eliminates the "blocking" problem, since only the reference set is "visible" to the vector being updated. More formally:

Let V be a feature vector in the original N-dimensional coordinate system of the feature space:

$$V = (a_1, a_2, \ldots, a_N)$$

Let F be a frame (spanning set for the original space) consisting of the M vectors (not necessarily linearly independent nor mutually orthogonal):

$$F = \{f_1, f_2, \ldots, f_M\}$$

$$\text{where } f_j = (b_{j1}, b_{j2}, \ldots, b_{jN}), \ j = 1 \text{ to } M$$

We define the Frame Vector for V to be the vector of distances:

$$D(V) = (d_1(V), d_2(V), \ldots, d_N(V))$$

where $d_J(V) = \|V - f_j\|$, and $\|.\|$ is a norm for the feature space.

NB: the metric in the lower-dimensional space need not be the metric in the original space. Optimization can still be carried out.

4   The Use of Torgerson Coordinates to Create "Super Features"

Torgerson Coordinates can be thought of as providing an alternate coordinate system for a data set. Given a distance matrix, the field-theoretic approach may be applied to recast a data set as vectors of Torgerson Coordinates in a space of any dimension, lower, equal to, or higher than that of the original feature space.

In another interpretation, the recasting described here can be thought of as an instance of the Vapnik "kernel trick", but one which is informed by the data distribution rather than merely an ad hoc transform such as a naively chosen polynomial.

This recasting can be optimized to preserve, to the extent possible, the spatial structure of the original data set in the target space, possibly under a norm differing from that in the original space.

Allowing any norm to be used in the target space opens many possibilities for creating alternative structure-preserving views of the original data.

The fact that Torgerson coordinates provide a lower dimensional proxy set than the original set suggests that, in some sense, the information they bear has been "concentrated" into a more compact representation. For this reason, we refer to the features in such a proxy set as super features.

These super features can be introduced back into the original data, or used to form entirely new feature sets for a problem. It will be seen that, in some applications, this intentional "rescaling redundancy" makes information more accessible (presumably by virtue of making it redundant).

This process may be applied to the original data set to create multiple projections, whose Cartesian product(s) constitute new feature sets having appended super features within which analysis of the original problem can be carried out. In particular, non-linearly separable problems can sometimes be greatly simplified in a new representation, as will now be described for three problems: the Nested Shells Problem, and the Parity-N Problem, and the BOT Detection Problem.

4.1   The Nested Shells Problem

A simple example is the nested shells problem in 3-dimensional space. For this problem, five spheres of different radii are centered at the origin. On each, 1000 points are randomly (uniform in each coordinate) distributed. The result is depicted in the figure at the right.

The colors indicate the "ground truth": blue points are on the innermost sphere, then green, cyan, red, and purple indicate points on the spheres giving classes 2 – 5, respectively. This provides an interesting, low-dimensional non-convex problem for testing classifiers.

Fig. 2. The Nested Shells Problem (Color figure online)

To produce a representation consisting entirely of super features, the set is reduced to a 2-dimensional set of Torgerson Coordinates, and also to a 1-dimensional set of

Torgerson Coordinates. The 1-D coordinates become feature 1, and the 2-D coordinates become features 2 and 3, respectively of a new representation of the data. The resulting 3-dimensional set can now be classified by a small number of linear decision surfaces (planes in this case), as the non-convexity has been eliminated. The new set is depicted in the figures below (Figs. 2 and 3).

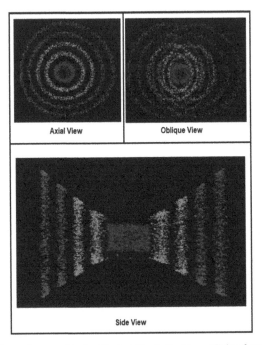

Fig. 3. Super features for the Nested Shells Problem (Color figure online)

4.2 The Parity-N Problem

The use of super features can be applied to a very difficult classification problem called the Parity-N Problem. It was the inability of a 1-layer perceptron to solve the parity-2 problem (of which much was made in the 1969 book titled <u>Perceptrons</u> by Minsky and Papert) that derailed U.S. government funding for neural network research for 10 years.

Consider all the unique strings consisting of N zeros and ones (there are 2^N such strings.). Geometrically, viewed as numeric vectors having coordinates 0 and 1, these data are the corners of the N-dimensional hypercube. The problem is to construct a classifier that ingests such a vector and determines whether its coordinates consist of an even or odd number of 1's… its parity.

This problem is interesting because it can be solved by an extremely simple rule, *assuming the defining nature of the data is known a priori.* But from the perspective of an agnostic classifier (like a neural network or N-Nearest-Neighbor machine), this problem is optimized for difficulty. In particular, every vector has N neighbors at the minimum distance 1 from it... and all are of the opposite class. For the parity problem, proximity *makes points different*, not similar. However, when the 1-dimensional Torgerson super feature is created for the Parity-10 problem (1,024 binary vectors), the classes become linearly separable (by a small number of points, in this case).

4.3 BOT Detection Application from Linguistic Analysis

The detection of automated agents (BOTS) in social media is a vexing problem which daily affects millions of individuals (e.g., by distributing SPAM, implementing fraud, and causing network congestion). It also has economic implications in advertising, entertainment, and politics, where the number of "Friends", "Followers", and "other contacts is the basis for decisions about strategic resource allocation.

The authors have formulated a linguistic mapping for Twitter accounts by assigning numeric codes to 127 topics that frequently appear in social media. A Twitter account can then be profiled by compiling a histogram of topical term occurrence by the user of that account. This encodes an account's history of linguistic discourse as a numeric vector which can be applied during BOT detection.

Field-theoretic super features have utility in this application, as indicated by the following experiment.

A corpus of 8,840 Twitter accounts was profiled using the coding method described above (topic histogramming). Ground truth BOT/notBOT assignment was provided for each account by an Expert System created for this purpose.

A set of ten super features was generated, and a set of 2 super features was generated. The sets were split evenly into disjoint Training and Test sets, each consisting of 4,420 vectors.

The two super features were prepended (as new features 1 and 2) to a copy of the 10 super feature set to obtain a set having 12 super features.

Visually, the two sets look very similar, as seen in a frame from a high-dimensional dynamic visualization below. The 10 super feature set is in the left-hand frame, and the 12 super feature set is in the right-hand frame. Blue points are normal accounts, while green points are BOT accounts. A large number of BOT accounts are embedded in the central main body, making this a difficult problem (Fig. 4):

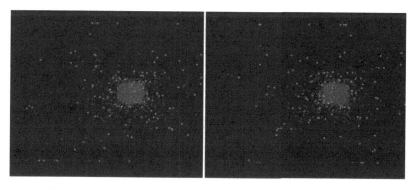

Fig. 4. 10-dimensional and 12-dimensional super features visualized

This allowed two experiments:

- An "Adaptive Bayesian" classifier was trained on the 10 super feature training set, and tested on the 10 super feature Test set.
- An "Adaptive Bayesian" classifier was trained on the 12 super feature training set, and tested on the 12 super feature Test set.

The confusion matrices for the two experiments are below. While the two additional super features had virtually no effect on the classifiers' performance on the BOT class (class 2), they significantly reduced the number of false BOT detections (from 156 to 77).

Overall Accuracy is the percent of the vectors correctly classified.
Geometric Accuracy is the geometric mean of the class precisions.
Geometric Recall is the Geometric mean of the class recalls (Fig. 5).

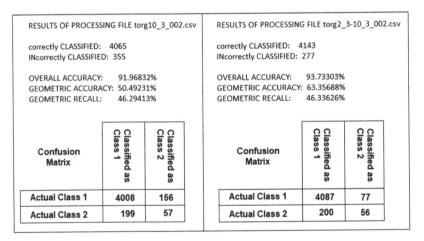

Fig. 5. Confusion matrices for the BOT Detection Problem

This result is typical of what has been observed in similar experiments on other real-world problems. This fact, along with the unexpected positive impact seen in the Nested Shells and Parity-N problems suggests additional work in this area is warranted.

References

1. Hancock, M., et al.: Field-theoretic modeling method for emotional context in social media: theory and case study. In: Proceedings of the 17th International Conference on Human-Computer Interaction, Los Angeles, CA, USA (2015)

2. Hancock, M., et al.: Fusing syntax and semantics for topic clustering in social media. In: Proceedings of the 19th International Conference on Human-Computer Interaction, Vancouver, Canada (2017)
3. Torgerson, W.S.: Multidimensional scaling: I. Theory and method. Psyohometrika **17**, 401–419 (1952)
4. Minsky, M., Papert, S.: Perceptrons: an introduction to computational geometry (1969). ISBN 0 262 13043 2

Non-monotonic Bias-Based Reasoning Under Uncertainty

Monte Hancock[(⊠)]

4Digital, Inc., Los Angeles, USA
practicaldatamining@gmail.com

Abstract. Not all "facts" are created equal.

Every piece of evidence has a pedigree which includes the credibility of the source, means of transmission, timeliness, precision, topical relevance, method of interpretation, and effectiveness of conformation with other evidence. None of these elements comes with a guarantee of perfect fidelity, so evidence-based reasoning is inherently uncertain in the sense that even valid reasoning can produce **untrue** conclusions.

Monotonic reasoning strategies are those assuming that the truth values of "facts" do not change during the reasoning process. In actual practice "facts" may be revised, updated, or even contradicted as evidence is collected. These changes can require the withdrawal of previous assertions and amendment of subsequently drawn conclusions. Reasoning strategies supporting both the assertion and withdrawal of facts are said to be non-monotonic.

Reasoning under uncertainty occurs when asserted "facts" are imprecise, in-commensurable, incomplete, and/or false. Imprecision is inherent in all measurement error, but the more important aspect of uncertainty is the presence of outright misrepresentation of information, whether accidentally through error, or intentionally by an adversary.

This presents a simple reasoning theory in stages as incrementally improved thought models.

Keywords: Rule-based system · Machine learning · Trainable expert system · Non-monotonic reasoning · Reasoning under uncertainty

1 Requirements for Reasoning

1.1 Human Reasoning

While propositional and predicate logic are powerful reasoning tools, they do not mirror what human experts actually do… nor do decision trees, Bayesian analysis, neural networks, or support vector machines.

Pose a problem for a human expert in their domain, and you will find, even given no evidence, that they have an a priori collection of beliefs about the correct conclusion. For example, a mechanic arriving at the repair shop on Tuesday morning already holds certain beliefs about the car waiting in Bay 3 before she knows anything about it. As

© Springer Nature Switzerland AG 2020
D. D. Schmorrow and C. M. Fidopiastis (Eds.): HCII 2020, LNAI 12196, pp. 250–265, 2020.
https://doi.org/10.1007/978-3-030-50353-6_19

she examines the car, she will update her prior beliefs, accruing "bias" for and against certain explanations for the vehicle's problem. At the end of her initial analysis, there will be some favored (belief = large) conclusions, which she will test, and so accrue more belief and disbelief. Without running decision trees, applying Bayes' Theorem, or using margin maximizing hyperplanes, she will ultimately adopt the conclusion she most believes is true. It is this "preponderance of the evidence" approach that best describes how human experts actually reason, and it is this approach we seek to model.

Bias-based Reasoning (BBR) is a proprietary mathematical method for automating implementation of a belief-accrual approach to expert problem solving. It enjoys the same advantages human experts derive from this approach; in particular, it supports automated learning, conclusion justification, confidence estimation, and natural means for handling both non-monotonicity and uncertainty.

Dempster-Shafer Reasoning is an earlier attempt to implement belief-accrual reasoning, but suffers some well-known defects [Lotfi paradox, constant updating of parameters, monotonic, no explicit means for uncertainty]. BBR overcomes these.

1.2 Using Facts in Rules

For simplicity and definiteness, the reasoning problem will be described here as the use of evidence to select one or more possible conclusions from a closed, finite list that has been specified a priori (the "Classifier Problem").

Expert reasoning is based upon facts (colloquially, "interpretations of the collected data"). Facts function both as indicators and contra-indicators for conclusions. Positive facts are those that increase our beliefs in certain conclusions. Negative facts are probably best understood as being exculpatory: they impose constraints upon the space of conclusions, militating against those unlikely to be correct. Facts are salient to the extent that they increase belief in the "truth", and/or increase "disbelief" in untruth.

A rule is an operator that uses facts to update beliefs by applying biases. In software, rules are often represented as structured constructs such as IF-THEN-ELSE, CASE, or SWITCH statements. We use the IF-THEN-ELSE in what follows.

Rules consist of an antecedent and a multi-part body. The antecedent evaluates a BOOLEAN expression; depending upon the truth-value of the antecedent, different parts of the rule body are executed.

The following is a notional example of a rule. It tells us qualitatively how an expert might alter her beliefs about an unknown animal should she determine whether or not it is a land-dwelling omnivore:

```
If  (habitat = land) AND (diet = omnivorous) THEN
    INCREASE BELIEF(primates, bugs, birds)
    INCREASE DISBELIEF(bacteria, fishes)
ELSE
    INCREASE DISBELIEF(primates, bugs, birds)
    INCREASE BELIEF(bacteria, fishes)
End Rule
```

If we have an INCREASE BELIEF function, and a DECREASE BELIEF function ("aggregation functions", called AGG below), many such rules can be efficiently implemented in a looping structure:

In a data store:

Tj($\mathbf{F}$i) truth-value of predicate j applied to fact $\mathbf{F}$i
bias(k, j, 1) belief to accrue in conclusion k when predicate j true
bias(k, j, 2) disbelief to accrue in conclusion k when predicate j is true
bias(k, j, 3) belief to accrue in conclusion k when predicate j false
bias(k, j, 4) disbelief to accrue in conclusion k when predicate j is false
Multiple rule execution in a loop:

```
IF Tj(F)=1 THEN                           'if predicate j true for Fi...
FOR k=1 TO K                              'for conclusion k:
Belief (k)=AGG(B(k,i),bias(k,j,1))        'true:accrue belief bias(k,j,1)
Disbelief(k)=AGG(D(k,i),bias(k,j,2))      'true:accrue disbelief bias(k,j,2)
NEXT k
ELSE
FOR k=1 TO K                              'for conclusion k:
Belief(k)=AGG(D(k,i),bias(k,j,3))         'false: accrue belief bias(k,j,3)
Disbelief (k)=AGG(B(k,i),bias(k,j,4))'false:accrue disbelief bias(k,j,4)
NEXT k
END IF
```

This creates a vector B of beliefs (b(1), b(2), ..., b(K)) for each of the conclusions 1, 2, ..., K, and a vector D of disbeliefs (d(1), d(2), ..., d(K)) for each of the conclusions 1, 2, ..., K. These must now be adjudicated for a final decision.

Clearly, the inferential power here is not in the rule structure, but in the "knowledge" held numerically in the biases. As is typical with heuristic reasoners, BBR allows the complete separation of knowledge from the inferencing process. This means that the structure can be retrained, even repurposed to another problem domain, by modifying only data; the inference engine need not be changed. An additional benefit of this separability is that the engine can be maintained openly apart from sensitive data.

Summarizing (thinking again in terms of the Classifier Problem):

When a positive belief heuristic fires, it accrues a bias $\beta > 0$ that a certain class is the correct answer; when a negative heuristic fires, it accrues a bias $\delta > 0$ that a certain class is *not* the correct answer. The combined positive and negative biases for an answer constitute that answer's belief.

After applying a set of rules to a collection of facts, beliefs and disbeliefs will have been accrued for each possible conclusion (classification decision). This ordered list of beliefs is a belief vector. The final decision is made by examining this vector of beliefs, for example, by selecting the class having the largest belief-disbelief difference (but we will formulate a better adjudication scheme below).

1.3 Problems and Properties

There are two major problems to be solved; these are, in a certain sense, "inverses" of each other:

The Adjudication Problem: reasoning forward from biases to truth

- What is the proper algorithm for combining accrued positive and negative biases into an aggregate <u>belief vector</u> so that a decision can be made?

The Learning Problem (reasoning backwards from truth to biases)

- Given a collection of heuristics and tagged examples, how can the bias values to accrue, β_{kl} and δ_{jl}, be determined?

Conventional parametric methods (e.g., Bayesian Inferencing), compute class likelihoods, but generally do not explicitly model "negative" evidence. Rather, they increase likelihoods for competing answers. They are inherently "batch" algorithms, performing their analysis after all evidence has been presented. They have the nice characteristic that they are capable of directly modeling the entire joint distribution (though this is rarely practical in actual practice). Their outputs are usually direct estimates of class probabilities.

Bias-based Reasoning (BBR) does not model the entire joint-distribution, but begins with the assumption that all facts are independent. This assumption is generally false for the entire population. We have found that this is effectively handled by segmenting the population data into strata within which independence holds approximately; rules are conditioned to operate within particular strata.

BBR supports both batch and incremental modes. It can "roll up" its beliefs after all evidence has been collected, or it can use an incremental aggregation rule to adjusts its bias with respect to each class as evidence is obtained.

1.4 Desirable Properties for a BBR

- Final conclusions should be independent of the order in which the evidence is considered.
- The aggregation rule should have compact range, e.g., it must have no gaps, and there must be a maximum and minimum bias possible.
- A bias of zero should mean that evidence for and against an answer are equal.
- <u>Symmetric Non-monotonic</u> reasoning should be supported, that is, it should be possible to withdraw facts previously asserted and obtain the same result as if these facts had never been considered. (<u>symmetry</u>: Accruing an amount of belief/disbelief and then accruing the same amount of disbelief/belief should leave the aggregate belief unchanged.)

2 Transforming Evidence into Beliefs Monotonically, and with No Provision for Uncertainty

2.1 Aggregation Rules

We first look at a simple aggregation rule that has many of the desired properties for BBR, but does not handle monotonicity or reasoning under uncertainty. We begin with this because it is easy to understand, and captures the essence of the reasoning theory. Also, it produces excellent results as a classifier when truth is sure and stable (often true for retrospective situations such as credit scoring and clinical diagnosis). In Sect. 3, it will be generalized for monotonicity and reasoning under uncertainty.

When rules fire in a bias-based KBES, they cause belief and/or disbelief in hypotheses to be accrued. A rule r is said to have isolated belief b if b is the total belief when rule r fires, and no other rules fire. (NB: Beliefs b are always in [0, 1]).

Proper aggregation of belief is essential: belief is not naively additive. For example, if I have 20 pieces of very weak evidence, each giving me 5% confidence in some conclusion, it would be foolish to assert that I am 100% certain of this conclusion just because $20 \times 5\% = 100\%$.

We proceed as follows: Consider two rules, r_1 and r_2, having isolated beliefs b_1 and b_2, respectively. What shall be the aggregate belief when both fire? We define the simple aggregation rule for this two-rule system as the "probability AND":

$$\text{aggregate belief}\,(r_1\,\text{and}\,r_2) = b_1 + b_2(1 - b_1) = b_1 + b_2 - b_1 b_2$$
$$= 1 - (1 - b_1)(1 - b_2)$$

Here if I have two pieces of evidence, each giving me 50% belief in a conclusion, after aggregation using this rule my belief in this conclusion is:

$$1 - (1 - 0.5)(1 - 0.5) = 0.75 = 75\%$$

(Using this rule for the case of twenty 5% beliefs, we arrive at about 64% belief, far from certainty!)

The simple aggregation rule says that we accrue additional belief as a fraction of the "unused" belief.

If a third rule r_3 with isolated belief r_3 fires, we find:

$$\text{aggregate belief}\,((r_1\,\text{and}\,r_2)\,\text{and}\,r_3) = (b_1 + b_2(1 - b_1)) + b_3(1 - (b_1 + b_2(1 - b_1)))$$
$$= b_1 + b_2 + b_3 - b_1 b_2 - b_1 b_3 - b_2 b_3 + b_1 b_2 b_3$$
$$= 1 - (1 - b_1)(1 - b_2)(1 - b_3)$$

In general, firing J rules r_j having isolated beliefs b_j gives aggregate belief:

$$\text{aggregate belief}\,(r_1\,\text{AND}\,r_2\,\text{AND}\,\ldots\,\text{AND}\,r_J) = 1 - \prod_j \{1 - b_j\} \quad (*)$$

The aggregate belief can be accumulated by application of the simple aggregation rule as rules fire. For, if $J-1$ rules have fired, giving a current belief of $\underline{b}$, and another rule r_J having isolated belief b_J fires, the simple aggregation rule gives a new belief of $\underline{b} + b_J(1 - \underline{b})$, which is easily shown to be in agreement with $(*)$.

The simple aggregation rule is clearly independent of the order of rule firings, assumes values in [0, 1], and has partial derivatives of all orders in the b_j. In fact,

$$\mathbf{d}(1 - \prod_j \{1 - b_j\})/\mathbf{d}b_k = \prod_j \{1 - b_j\}/(1 - b_k)$$

Hence, all partials having multi-indices with repeated terms are zero.

Note that if any rule r_j has isolated belief $b_j = 1$, then the aggregate belief is 1; and, accruing an isolated belief of $b_j = 0$ has no effect on the aggregate belief.

2.2 The Adjudication Rule: Combining Positive and Negative Beliefs

We use the term Adjudication Rule (AR) for any method of combining a vector of positive beliefs with a vector of negative beliefs to obtain a belief vector from which a decision can be made by selecting the class having the greatest belief. This final belief vector is said to be adjudicated.

One simplistic adjudication rule would be to just subtract the negative beliefs from the positive beliefs. Keep in mind that the negative beliefs are non-negative values expressing the amount of disbelief in each hypothesis: both B and D assume values in $[0, 1]$.

Therefore, let:

1) $B(k, i)$, the belief that $H_k(\mathbf{F}_i)$ is true
2) $D(k, i)$, the disbelief that $H_k(\mathbf{F}_i)$ is true

Differencing these, we obtain a confidence that $H_k(\mathbf{F}_i)$ is true:

$$C(k, i) = B(k, i) - D(k, i)$$

This difference assumes values in $[-1, 1]$. The adjudicated confidence vector $\mathbf{C}(\mathbf{F}_i)$ has components $C(k, i)$:

$$\mathbf{C}(\mathbf{F}_i) = (B(1, i) - D(1, i), B(2, i) - D(2, i), \ldots, B(K,i) - D(K, i))$$

Since real multiplication is commutative, the value of $C(k,i)$ is independent of the order of application of the rules.

HOWEVER: no consideration was given in the formulation of this rule to either monotonicity or uncertainty. In particular, the aggregation and adjudication methods described so far are inherently non-monotonic: accruing a positive bias of b for a hypothesis and then accruing a negative bias of b for that hypothesis will not, in general, return the system to its original belief state. The methods are not "stateless". In Sect. 3, this shortcoming will be removed.

2.3 Computing Knowledge Directly

The "knowledge" of a parametric-rule KBES is in two places:

1) Q_{kjm}, which determine how much rules may modify beliefs
2) T_j, the predicate truth value functions, which can have parameters of their own.

We fix the T_j, and use a numerical technique to optimize the Q_{kjm} wrt a training set, i.e., in a supervised mode. (For example, a bisection technique may be used, avoiding the practical problems associated with direct gradient techniques).

The bias-based KBES described so far may be regarded as a function which maps feature space into the "confidence hypercube", the K-fold cartesian product of the interval $[-1, 1]$ with itself. Vectors in the confidence hypercube have as their components the confidences that an input feature vector is consistent with the respective K hypotheses (goal classes).

Given this geometric interpretation of a KBES, it is natural to use Euclidean distance in K-space to define the error function for rule optimization.

In supervised learning, there is supplied with each feature vector $\mathbf{F}_i$ in the training set a target confidence vector, $\mathbf{G}(\mathbf{F}_i)$, having components G(k, i). For a particular training set, the error function will be the sum of the Euclidean distances of the KBES outputs from the target confidence vectors.

3 Non-monotonic Reasoning Under Uncertainty

3.1 Summary of Bias-Based Reasoning Without Non-monotonicity

As we have seen, when there are many classes and many rules, the simple aggregation rule may be written as follows: Suppose that N belief heuristics and M disbelief heuristics have fired. The aggregate belief and disbelief for class k is:

$$B_k(\beta_{1k}, \ldots, \beta_{Nk}) = 1 - \prod_{i=1}^{N} (1 - \beta_{ik})$$

$$D_k(\delta_{1k}, \ldots, \delta_{Mk}) = 1 - \prod_{j=1}^{M} 1 - \delta_{jk}$$

We noted above that an aggregated belief vector could be formed by looking at excess belief over disbelief by differencing:

$$C(\beta_{1k}, \beta_{2k}, \ldots, \beta_{Nk}, \delta_{1k}, \delta_{2k}, \ldots, \delta_{Mk}) = (B_1 - D_1, B_2 - D_2, \ldots, B_L - D_L)$$

The variable C is used to indicate that these differences are not class probabilities, but class "confidences" having no particular normalization properties.

The properties of C are all attributes of the aggregation rule, the algorithm for combining positive and negative biases into a final L-dimensional vector of the beliefs for the L classes.

These aggregation formulae having the following properties:

1. They can be computed in a "batch method" as above, but they can both also be accumulated incrementally as evidence is gathered by:

$$newvalue = oldvalue + bias * (1 - oldvalue)$$

This allows individual rules to accrete belief, so that the computation of the belief vector is a side-effect of the rule firing process.
2. This rule has compact range = [0, 1].
3. This rule is C-infinity (partial derivatives are computed below)
4. If a bias of 0 is accrued, the old belief is not changed; if a bias of 1 is accrued the belief is 1, and will never decrease.
5. It provides no natural mechanism for handling uncertainty.

3.2 Aggregating with Non-monotonicity

First we introduce the mechanism for non-monotonicity. Suppose that N belief heuristics having magnitudes $\beta_{kl} \geq 0$ and M disbelief heuristics having magnitudes $\delta_{jl} \geq 0$ have been applied. Rather than differencing belief and disbelief, we use a quotient in the accrual: The aggregated bias with respect to class l is defined:

$$
\mathrm{B}l(\beta_{1,l}, \ldots, \beta_{N,l}, \delta_{1,l}, \ldots, \delta_{M,l}) = 1 - \left(\frac{\prod_{k=1}^{N} (1 - \beta_{kl})}{\prod_{j=1}^{M} (1 - \delta_{jl})} \right)
$$

Then the aggregated belief vector is:

$$
\vec{\mathrm{B}}(\beta_{1,l}, \ldots, \beta_{N,l}, \delta_{1,l}, \ldots \delta_{N,l}) = (\mathrm{B}_1, \ldots, \mathrm{B}_L)
$$

As before, B_k is our adjudicated belief in hypothesis k.

This expression for $\vec{\mathrm{B}}$ is clearly symmetric in its arguments, so belief and disbelief can be accrued in any order. Further, accruing a belief of b followed by accrual of a disbelief of b will place a $(1-b)$ term in both numerator and denominator of the quotient, which will divide out; this allows changes to belief from previous fact to be backed-out.

3.3 Model Development and Training

How do we determine the amount of belief, β_{kl}, and the amount of disbelief, δ_{jl}, to accrue for each class in each heuristic?

The components of $\vec{\mathrm{B}}$ are differentiable with respect to the beliefs and disbeliefs, so the gradient may be computed:

$$
\partial \mathrm{B}_l / \partial \beta_{\widehat{k}l} = \left(\frac{1}{1 - \beta_{\widehat{k}l}} \right) \left(\frac{\prod_{k=1}^{N} (1 - \beta_{kl})}{\prod_{j=1}^{M} (1 - \delta_{jl})} \right)
$$

$$
\partial \mathrm{B}_l / \partial \delta_{\widehat{j}l} = \left(\frac{-1}{1 - \delta_{\widehat{j}l}} \right) \left(\frac{\prod_{k=1}^{N} (1 - \beta_{kl})}{\prod_{j=1}^{M} (1 - \delta_{jl})} \right)
$$

These can be compactly rewritten:

$$
\partial \mathrm{B}_l / \partial \beta_{\widehat{k}l} = \frac{1 - \mathrm{B}_l}{1 - \beta_{\widehat{k}l}}
$$

$$
\partial \mathrm{B}_l / \partial \delta_{\widehat{j}l} = \frac{1 - \mathrm{B}_l}{1 - \delta_{\widehat{j}l}}
$$

These might make feasible optimization by a gradient technique, e.g., the Method of Steepest Descent:

$$
\beta_{kl} = \beta_{kl} - h_{1kl}(\partial \mathrm{B}_l / \partial \beta_{kl}), \; h_{1kl} \in \Re^+
$$

$$\delta_{jl} = \delta_{jl} - h_{2kl}(\partial B_l/\partial \beta_{jl}), \; h_{2kl} \in \Re^+$$

Given a set of data having ground truth tags, an iterative cycle such as that depicted below using this (or a similar) update rule can be used to <u>learn</u> the beliefs and disbeliefs that will cause the heuristics to give correct answers on the training set:

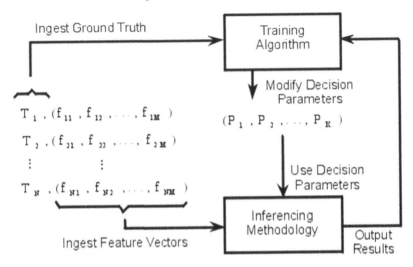

3.4 Finally: Non-monotonic Bias-Based Reasoning Under Uncertainty

Taking as a starting point the bias-based non-monotonic theory of the previous section, we finally introduce the mechanism for accommodating uncertainty. The goals are:

1) use an estimate of the level of certainty to adjust the impact a fact can have on beliefs
2) provide a means for retrospectively changing the uncertainties of facts already applied, and updating current beliefs accordingly

Uncertainty is an attribute of both the data source and the data itself ("good" sources sometimes provide bad data, for example). This means that our uncertainty mechanism must allow the assignment of "credibility" at the data-source level, and also to individual data from whatever source.

All of the nice properties of the previous aggregation and adjudication rules are preserved (with an one exception), and both non-monotonicity and accommodation of dynamic uncertainty are accommodated by the batch algorithm:

Suppose that there are M possible conclusions for the reasoner (e.g., M classes to which an entity is to be assigned). Let facts be interpreted and mapped to ordered 4-tuples, which we call <u>fact vectors</u>:

$$F_k = (s_k, t_k, c_k, V_k)$$

Here the kth fact F_k is from source s_k, with date-time-group t_k, a level of uncertainty, and a corresponding vector of features (=data measurements).

A Fact Base is a set of fact vectors $\mathbf{F} = \{F_1, F_2, \ldots, F_L\}$. The fact base is a catalog of what we have been told by our sources with variable certainty at various times.

A source is the originator of a fact. It could be a data collection system, data base, algorithm, manual estimate, etc.

Each fact has a time t_k. This is provided to give recency and latency information, but also so that facts with particular times can have their certainties updated as a group. For example, it might be useful to retrospectively change the certainty level of all facts from a given source for given times, updating current beliefs accordingly.

Each fact has a certainty c_k at time t_k. This is a numeric value between 0.0 and 1.0 inclusive. Facts that are "completely uncertain" have $c = 0$. Facts that are "completely certain" have $c = 1$. (In actual practice, these extreme values are unrealistic.) Intermediate levels of uncertainty fall between 0 and 1, with larger values indicating higher levels of confidence that the fact is actually true.

The input to a KBES is a fact base. The Inference Engine selects facts to be processed and rules to fire so that beliefs are updated in accordance with the IE's inferencing strategy.

To fire, a rule invoked by the IE evaluates its predicate at the given fact and apply its biases $(\beta_{j1}, \beta_{j2}, \ldots, \beta_{jM})$, thereby adjusting beliefs in the some or all of the possible conclusions $1, 2, \ldots j, \ldots, M$. The biases are applied using the aggregation rule.

3.5 The Comprehensive Aggregation Rule

The finalized theory demands symmetry: biases can be either positive or negative. Positive biases increase belief, while negative biases decrease belief. Two additional definitions are needed:

1. The Sgn Function (SIGNUM, SIGN, or SGN function) returns the algebraic sign of a real number:

$$Sgn(x) = -1 \text{ if } x \text{ is negative, otherwise } + 1$$

2. γ_s is the "gain" of source s. It is a value in $[0, 1]$ that is applied to all facts from source s, IN ADDITION TO the confidences c_j of facts from source s. It provides a source-specific "credibility tuning" value that will generally be set to 1. Possible applications will be mentioned below.

The finalized theory places a number of functional requirements specifically related to uncertainty on the aggregation rule:

U1. Facts firing with $c = 1$ provide the full update effect of their biases, β_{jk}
U2. Facts firing with $c = 0$ have no effect on beliefs
U3. Facts firing with intermediate values of c have intermediate effect, with larger c values having greater effect
U4. Facts from sources having $\gamma = 1$ provide the full update effect of their biases including the application of their c values.
U5. Facts from sources firing with $\gamma = 0$ have no effect on beliefs

U6. Facts from sources firing with intermediate values of γ have intermediate effect (including the application of their c values), with larger γ values allowing greater effect

In summary, c mitigates the impact of individual facts, while γ mitigates the impact of all facts from a given source.

We are now in a position to state the NMBBRUU Aggregation Rule for batch mode (all biases for each class have been determined):

Suppose that M belief heuristics for a particular conclusion k have been applied. Some of the biases β_{kj} are positive and some are not. Further, suppose that the confidence associated with the kth fact is c_k, and the gain of the source for the kth fact is γ_{sk}. The aggregated bias with respect to class k for the fact base is defined:

$$B_k = 1 - \prod_{j=1}^{N} \left[\frac{1 - |\beta_{kj}|}{1 - \gamma_{s_k}(1 - c_k)|\beta_{kj}|} \right]^{\text{sgn}(\beta_{kj})}$$

Then the aggregated belief vector is:

$$\vec{B} = (B_1, \ldots, B_L)$$

As before, B_k is our adjudicated belief in hypothesis k. Note that the purpose of the sgn function is to put negative beliefs into the denominator so they can "cancel" positive beliefs.

This incremental aggregation function satisfies all of the functional requirements for uncertainty processing.

To back out a previously assigned fact having uncertainty c_s and update current beliefs with revised uncertainty $\tilde{c}_s$, apply the following correction:

$$NewB = (1 - OldB)\left(\frac{1 - \gamma_k(1 - c_s)|\beta_k|}{1 - \gamma_k(1 - \tilde{c}_s)|\beta_k|} \right)^{\text{sgn}\,\beta_k}$$

This changes the current belief to what it would have been had fact k been fired with updated uncertainty $\tilde{c}_s$.

3.6 Additional Closing Thoughts

Many useful capabilities can be provided using the approach described here, including:

- the creation of confidence factors for conclusions
- the automatic generation of conclusion justification reports that explain the reasoners conclusions in natural domain terms
- automatic trainability (offline learning).

A particularly provocative application is the evaluation of the relative utility of information sources. It is possible that this could enable the detection of compromised sources (those that are providing low-quality information, or disinformation). This capability uses the source gain γ_s. By measuring the source-by-source sensitivity of reasoning quality, sources whose facts degrade overall performance can be identified. If "turning down the gain" on a source reduces overall system error, that source should be scrutinized.

Source gain sensitivity could provide a systematic method for identifying bad input, whether it arises from a bad system input, or a trusted source that has been compromised.

Source gain sensitivity could provide a systematic method for identifying bad input, whether it arises from a bad system input, or a trusted source that has been compromised.

4 Case Studies

4.1 Case Study 1: Jury (Excerpt from [2])

For this case study, bias-based reasoning was used to model of a specific cognitive behavior that brings two fundamental elements of human cognition into direct conflict: the purely emotional (referred to here as EMOS), and the purely rational (referred to here as NOOS). Psychologists refer to the mental experience associated with this conflict as cognitive dissonance.

We selected a jury in a criminal trial as the domain for experiments with this system, the task to be performed is to consider facts-in-evidence, and apply the rule bases to produce verdicts of "guilty", "not guilty", or "deadlocked". This scenario offers a fundamentally binary decision problem that can be easily understood without special knowledge. To make interpretation of the machine's decision processes transparent, EMOS and NOOS were developed with a conclusion justification capability by which each can express (in natural language) how the pieces of evidence affected their decisions.

"Cognitive dissonance refers to a situation involving conflicting attitudes, beliefs or behaviors. This produces a feeling of mental discomfort leading to an alteration in one of the attitudes, beliefs or behaviors to reduce the discomfort and restore balance." –Saul Mcleod, University of Manchester, Division of Neuroscience & Experimental Psychology [1].

A "double-minded" rule-based system was developed consisting of the two components EMOS and NOOS above. These two system components use the same decision-making algorithm, but different heuristics: to assess and combine facts, EMOS applies "soft emotional" factors, while NOOS applies "rigid principled" factors.

The outcomes and interpretation of the results of 255 "jury trials" were described. The figure below catalogs the "evidentiary analysis" of the NOOS jury member for a particular case.

Apply NOOS rules to the evidence. Each fact (A - H at right) can be TRUE or FALSE (values in column 1).

The next-to-last column shows new Aggregate Beliefs as facts are applied.

The last column is the message NOOS will append to its explanation report

A The Defendant has a criminal history.
B The Defendant has been identified as the perpetrator by eye witnesses.
C There is forensic evidence that ties the Defendant to the crime.
D The Defendant had a motive for the crime.
E The Defendant is a member of a minority racial or religious group.
F The Defendant is under 26 years of age.
G The Defendant has a history of gang membership.
H The Defendant dropped out of high school.

	EVIDENCE	If TRUE increase BEL, -ief by	If TRUE increase DIS-belief by	If FALSE increase BEL, -ief by	If FALSE increase DIS-belief by	Resulting BEL, -ief	Resulting DIS-belief	Resulting Aggregate Belief	JUSTIFICATION FOR ADJUSTMENT:
NOOS ADJUST GUILTY BELIEF					Prior = 0.60				
TRUE	A - HISTORY	0.014	0	0	0	0.6056	0.4591497	0.2056	If they broke the law before they are likely to break it again.
FALSE	B- WITNESS	0	0	0	0.008	0.6056	0.4048	0.2008	Eye witnesses say what they are told to say.
FALSE	C-FORENSICS	0	0	0	0.036	0.6056	0.4262272	0.1793728	Police labs are known to be careless and unreliable.
TRUE	D-MOTIVE	0.012	0	0	0	0.6103328	0.4262272	0.1841056	Criminals don't need a good reason to commit a crime.
TRUE	E-MINORITY	0.004	0	0	0	0.6118914	0.4262272	0.1856643	The Defendant is a product of their environment.
FALSE	F-UNDER 26	0	0	0	0.012	0.6118914	0.4331124	0.178779	Ignorance is no excuse for bad behavior.
TRUE	G-GANG	0.006	0	0	0	0.6142201	0.4331124	0.1811076	Those who choose to join gangs are choosing criminality.
TRUE	H-DROPOUT	0.004	0	0	0	0.6157632	0.4331124	0.1826507	Laziness and poor work ethic are signs of bad character.
NOOS ADJUST INNOCENT BELIEF					Prior = 0.40				
TRUE	A - HISTORY	0.028	0	0	0	0.4168	0.4331124	-0.1832	If they broke the law before they are likely to break it again.
FALSE	B- WITNESS	0	0	0	0.012	0.4168	0.6048	-0.1880001	Eye witnesses say what they are told to say.
FALSE	C-FORENSICS	0	0	0	0.054	0.4168	0.6261408	-0.2093408	Police labs are known to be careless and unreliable.
TRUE	D-MOTIVE	0.024	0	0	0	0.4307968	0.6261408	-0.195344	Criminals don't need a good reason to commit a crime.
TRUE	E-MINORITY	0.008	0	0	0	0.4353504	0.6261408	-0.1907904	The Defendant is a product of their environment.
FALSE	F-UNDER 26	0	0	0	0.018	0.4353504	0.6328703	-0.1975199	Ignorance is no excuse for bad behavior.
TRUE	G-GANG	0.012	0	0	0	0.4441262	0.6328703	-0.1907441	Those who choose to join gangs are choosing criminality.
TRUE	H-DROPOUT	0.008	0	0	0	0.4465892	0.6328703	-0.1862811	Laziness and poor work ethic are signs of bad character.

ADJUDICATE NOOS

GUILTY belief: 0.6157632   GUILTY Disbelief: 0.4331124   NOOS GUILTY AGG: 0.1826507
INNOCENT belief: 0.4465892   INNOCENT Disbelief: 0.6328703   NOOS INNOCENT AGG: -0.1862811
NOOS DECISION: The Defendant is guilty.   Confidence = |0.1826507 - (-0.1862811)| = 0.3689318

4.2 Case Study 2: Parole Board (Excerpt from [3])

For this case study, a hypothetical population of parole candidates was created, having values for the following attributes:

f1: severity of the offense (1–5}
f2: Gender (1 = male, 2 = female)
f3: Age (0–100 years)
f4: Contrition (1–5, abstract scale, 1 = no contrition, 5 = high contrition)
f5: Offense_History (# previous incarcerations)
f6: Parole History (# previous Board appearances for this offense)
f7: Behavior (# serious disciplinary actions this incarceration)
f8: Mental Health (0.0–1.0, abstract scale, 0.0 = unhealthy, 1.0 = very healthy)
f9: % current sentence that has been served (0% to 100%)
f10: Strength of Active Family Support (0 = no effective support, 1 = effective support, abstract scale)
f11: Employability (0 = not employable, 1 = employable, abstract score)

Bias-based reasoning was used to consider and adjudicate a parole decision for each of the 255 possible combinations of candidate attributes. It is common practice for such decisions to be based in part on a scoring or point system. The bias-based reasoner makes it possible to specifically quantify how each factor affects the parole decision and applies these factors objectively and consistently. One of the heuristics in the reasoner follows. Notice that the rule accumulates its conclusion justification (cjr) as a side-effect of its operation. By producing the cjr accumulates from all the rules, a summary Conclusion Justification report is produced automatically. It can be annotated to indicate numerically how each factor impacted the decision.

```
Rule 10: Family Support System and Employability
'Fuses Two Features: f10="FamilySupport", f11="Employability"
rel_fea=10 'Strength of Active Family Support
rel_fea_2=11 'Employability
cjr=""
fvrel=fv(vector,rel_fea)
fvrel_2=fv(vector,rel_fea_2)
stat=fvrel 'statistic to cite
'
if fvrel=0 and fvrel_2=0 then
beta=0.0
delta=0.15
cjr="-This candidate has no family support system, and no employ-
able skills."
elseif fvrel=0 and fvrel_2=1 then
beta=0.05
delta=0.0
cjr="+This candidate has no family support system but does have
employable skills."
elseif fvrel=1 and fvrel_2=0 then
beta=0.1
delta=0.0
cjr="+This candidate has a family support system but has no employ-
able skills."
elseif fvrel=1 and fvrel_2=1 then
beta=0.25
delta=0.0
cjr="+This candidate has a family support system and has employable
skills."
end if
```

4.3   Case Study 3: Wellness Recommender System (Excerpt from [4])

A wellness recommender system has been created using bias-based reasoning to assess
user wellness and make recommendations to most effectively move the user to a wellness
state conditioned by their demographic and personal characteristics. The recommender
operates in two phases: Phase 1 assesses the user's wellness values in 25 areas covering
psychology, physiology, and finances. Information is collected/inferred from information
they provide, information collected from sources authorized by the user, and from one
or more wearable devices on the user's person.

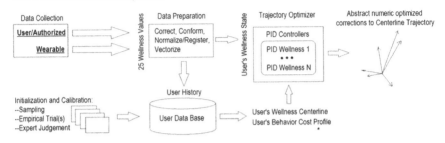

Phase 1. Mathematical Assessment and Optimization

Phase 2: Translation and Presentation

The first expert system applies bias-based reasoning to assess the user's wellness state. This assessment is ingested by the second expert system, which uses bias-based reasoning to formulate wellness recommendations (specific recommended behaviors). These are presented to the user in a summary report, generated automatically by the expert system in colloquial natural language suggesting actions to address the user's areas of greatest deviation from wellness for their demographic. Two compliance factors specific to the individual user are applied to select recommendations most likely to be adopted by the user.

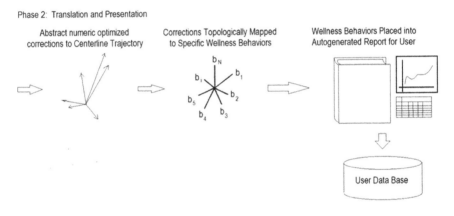

Phase 2: Translation and Presentation

References

1. Hancock, M.: Practical Data Mining, CRC Press (2012). ISBN 978-1-4398-6836-2
2. Hancock, M., et al.: Cognitive dissonance in a multi-mind automated decision system. In: Proceedings of the 21nd International Conference on Human-Computer Interaction, Las Vegas, NV, USA, July 2019
3. Hancock, K., et al.: Parole board personality and decision making using bias-based reasoning. In: Proceedings of the 20th International Conference on Human-Computer Interaction, Vancouver, Canada, July 2018
4. Hanlon, B, Hancock, M., et al.: Wellness optimization through feedback control. In: Proceeding of the 22nd International Conference on Human-Computer Interaction, Copenhagen, Denmark, July 2020
5. Hartney, C., Vuong, L.: Racial and Ethnic Disparities in the US Criminal Justice System. National Council on Crime and Delinquency, Oakland (2009)

6. Lipsey, M.W., Howell, J.C., Kelly, M.R., Chapman, G., Carver, D.: Improving the effectiveness of juvenile justice programs: A new perspective on evidence-based practice. Center for Juvenile Justice Reform, Georgetown Public Policy Institute (2010)

Explainable Artificial Intelligence: What Do You Need to Know?

Sam Hepenstal[1]([⊠]) and David McNeish[2]

[1] Defence Science and Technology Laboratory, Porton Down, Salisbury SP4 0JQ, UK
shepenstal@dstl.gov.uk
[2] Defence Science and Technology Laboratory, Portsdown West, Fareham PO17 6AD, UK
dgmcneish@dstl.gov.uk

Abstract. In domains which require high risk and high consequence decision making, such as defence and security, there is a clear requirement for artificial intelligence (AI) systems to be able to explain their reasoning. In this paper we examine what it means to provide explainable AI. We report on research findings to propose that explanations should be tailored, depending upon the role of the human interacting with the system and the individual system components, to reflect different needs. We demonstrate that a 'one-size-fits-all' explanation is insufficient to capture the complexity of needs. Thus, designing explainable AI systems involves careful consideration of context, and within that the nature of both the human and AI components.

Keywords: Explainable AI · Algorithmic transparency · Interpretability

1 Introduction

The use of artificial intelligence (AI) to support high risk and high consequence decision making faces a significant challenge; can the system outcomes and the processes by which they are reached be explained to the human decision maker? This is particularly relevant within the defence and security domains where decisions have profound consequences and may, for example, result in detention, injury or even death. Human accountability for these types of decisions is of paramount importance, but is dependent on the decision making process being intelligible to human agents (Floridi and Cowls 2019). Based on recent research by the authors, this paper seeks to address the following questions: What requires an explanation in order to achieve explainable AI? And how might explanation needs vary across different user roles or system components? This research focusses on the defence and security domain but the findings are likely to be generalizable to the use of AI to support high risk and high consequence decision making more broadly.

2 There Are Various Interpretations of Explainable AI

Explainable AI (XAI) is a slippery notion which can refer to several distinct concepts (Lipton 2016) and there is as yet no widely agreed definition. As pointed out by Preece

© Crown Copyright, Dstl 2020
D. D. Schmorrow and C. M. Fidopiastis (Eds.): HCII 2020, LNAI 12196, pp. 266–275, 2020.
https://doi.org/10.1007/978-3-030-50353-6_20

(2018) the desire to understand and explain the outputs from AI-based systems is not new, although recent developments in the field of Machine Learning (ML) have led to a resurgence in interest. Despite the abundance of overlapping terminology within the field of XAI, this section aims to describe some of the key concepts which have emerged from the literature.

Hoffman et al. (2018) describe the aim of XAI as helping human agents develop a strong mental model of the system in order to provide one or more of the following benefits; justification for individual outputs, an understanding of how the system works in general including its capabilities and limitations, and the ability to accurately predict what it would do given certain conditions. According to Hoffman and Klein (2017), from a psychological and philosophical perspective, explanation is closely related to causal reasoning and includes abduction, retrospection (including counterfactual reasoning) and prospection (projection into the future). Causal reasoning is central to sense-making, the development of mental models, decision making, re-planning, coordination and antici-patory thinking. In the broadest sense, XAI seeks to answer questions such as: What is the system doing? How is it doing it? Why is it doing it? And what would it do given a certain set of conditions? (Hoffman et al. 2018; Weller 2017).

An important distinction to draw is between local and global explanation. Local explanation focusses on justifying a single decision or output, whilst global explanation focusses on overall system behaviour (Hoffman et al. 2018; Weller 2017; Doshi-Velez and Kim 2017; Ribeiro et al. 2016). XAI is often described in terms of transparency and post-hoc explanation (Preece et al. 2018; Mittelstadt et al. 2019; Lipton 2016; Preece 2018). Transparency seeks to reveal information about the internal structure of a model and its training data in order to communicate how the system reaches an output (Tomsett et al. 2018, Preece2018). Post-hoc explanations aim to provide information about the cause or reason for an output (Tomsett et al. 2018) generally without referring to the internal structure of the model (Preece 2018). Or put another way, transparency focusses on how the system works internally whilst post-hoc explanations address how the model behaves (Mittelstadt et al. 2019). These ideas are echoed within the framework (Fig. 1)

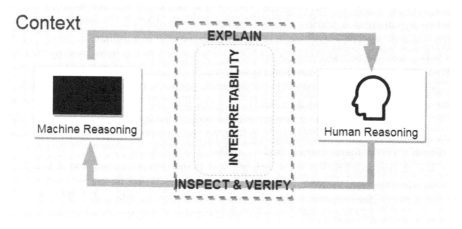

Fig. 1. Algorithmic transparency framework (Hepenstal et al. 2019).

proposed by Hepenstal et al. (2019) which distinguishes between the ease with which a user can explain the results provided by a system, and the ability to inspect and verify the goals and constraints of the system within context.

3   Roles Influence Explanation Needs

Ultimately explanation needs to be context specific; the goals, motivations, needs, capabilities and limitations of the target audience must be considered (Tomsett et al. 2018; Weller 2017; Ribeiro et al. 2016). Drawing on the division between expert and lay users proposed by Ras et al. (2018), as well as the simple distinctions made in early AI research (Preece 2018), the stakeholders who need XAI can be grouped into two broad categories; developers and users. As presented in Fig. 2, the terms developers and users are used by the authors to mean those who influence system design and those affected by the system respectively.

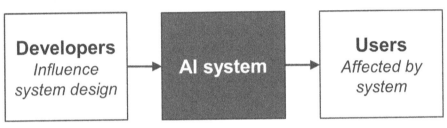

Fig. 2. Two broad categories of stakeholders requiring XAI.

Our research sought to investigate the explanation needs of these two stakeholder groups through a participatory design workshop. This involved twelve participants with either a military or AI background, from either the developer or user stakeholder communities. The twelve participants were split into three groups of four with a mixture of backgrounds in each to encourage interdisciplinary working. The workshop focussed on two hypothetical scenarios which were used to stimulate and focus discussion by illustrating the possible application of AI decision support in a military context. These scenarios were not intended to represent any in-service or future systems used by the UK MOD. The first scenario focussed on a system which analysed data from sensors worn by infantry soldiers in order to support task allocation decisions made by their commander. The second focussed on a system which analysed ship sensor information in order to detect and classify likely incoming threats and support the commander in choosing the best course of action. The workshop approach was based on the design thinking process and built upon the methods described by McNeish and Maguire (2019).

The participants felt that explainability was highly important for both developers (those who influence system design) and users (those affected by the system) of AI-based systems. Two reasons for its importance, common to both stakeholder groups, were regulatory and legal compliance and the ability to investigate when something goes wrong. Whilst developers need explainability for testing and iterative improvement of system performance, users need to build their trust and confidence in the system when optimising and being able to justify their decision making.

Five key stakeholder needs emerged and are summarised in Fig. 3. Each of these attributes is considered from the perspective of the two stakeholder groups. We draw out the differences, if any, and provide a rationale as to why that may be the case. Overall the most common theme was the need to understand the input data and sensor performance, with almost twice as many comments attributed to it than any other individual theme. The remaining four received almost the same number of comments as one another. These five areas of explanation needs are not unique to military applications of AI systems. However the diverse and dynamic nature of military operations, including the potential for lethal force, may drive their particular relevance within the military domain.

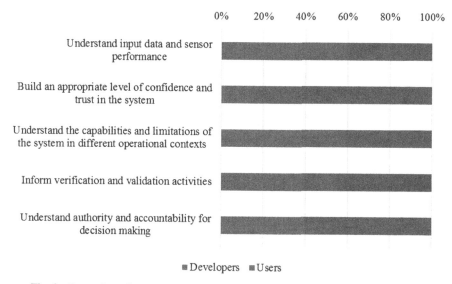

Fig. 3. Proportion of comments attributed to each stakeholder group for each theme.

Although the need to understand the input data and sensor performance was considered relevant to both stakeholder groups, it was attributed more often to developers. Both users and developers may need to know the number and type of sensors providing the data used as inputs to the system, whether or not these sensors are working, the impact of the operational environment on sensor reliability, and their operational range. Ultimately they are likely to want to know how all of these factors could impact the accuracy and reliability of the outputs from a decision support system. They also need to know, particularly from a developer's perspective, whether the right parameters are being measured; in other words how reliable the input metrics are as indicators for the task performance measures being predicted. Users in particular need to understand the 'completeness' of the input data being used as a basis for predictions including whether any particularly important sensors were not being used. From a developer perspective understanding the minimum number and type of information sources required to produce accurate and reliable predictions is important, particularly when used to inform safety critical decisions. When using multiple data sources as inputs, understanding the relative priority of each and its influence on prediction accuracy would be important for both developers and users, particularly if different data sources seem to contradict. In summary, both developers and users will need an understanding of the provenance of the data being used to train the system and the inputs used to inform real-time predictions.

Building an appropriate level of confidence and trust in the system was a challenge most frequently associated with the user stakeholder group. This included the level of confidence or uncertainty in the data sources and the system outputs and across different types of scenario. Both justification for individual outputs as well as an understanding of the system reliability across multiple uses were seen as necessary for informing this appropriate level of user confidence and trust.

Both developers and users need to understand the capabilities and limitations of the system when used in different operational contexts, and therefore how generalizable the system might be. Contextual factors could include different physical environments, missions and tasks, types of threat, and interoperability with other systems (for example during coalition operations). These factors may have implications for what training data should be used and the resultant prediction accuracy. Also related to this theme was the need to understand how the system is likely to respond to an unexpected or unknown threat.

The need to inform the verification and validation activities was a common theme across both scenarios. In particular this included understanding the sufficiency of test cases used during the verification of the system as well as the characteristics of the training and input data. Explanation may be required when validating the system against the user requirements; in particular, the ability to assess its impact on overall mission performance and the face validity of information being presented to the user. Another specific concern was understanding a system which adapts and learns over time; an online learning system for example. This may require some form of through-life assurance which would require the system to enable both developers and users to understand how it has changed and whether this might impact the system safety argument.

Finally, authority and accountability and was the most common theme within the second scenario which focussed on threat detection in a maritime environment. Both developers and users need to understand who is accountable for safety-critical and mission-critical decisions. Who is responsible for decisions and how much authority the human and the system should have over individual actions must be clear in order to inform design decisions relating to human-machine interaction, assessment by regulators, and to ensure the users understand what is expected of them. This is particularly important from a legal perspective and for the management and application of rules of engagement to ensure that there is sufficient human control over the ultimate outcome. Although the legal and ethical drivers for XAI are mentioned in previous research (Lipton 2016; Samek et al. 2017), the nature of military decision making which by definition could lead to the use of lethal force, adds significant weight to this challenge and has legal implications relating to international humanitarian law and the application of rules of engagement. Put simply the severe consequences of inappropriate levels of trust in a ML system drive the requirement for clarity over decision making roles.

4 System Components Influence Explanation Needs

Real world systems are complex, often incorporating many algorithms and components. Our research to date has considered a conversational agent (CA) system for information retrieval, to support criminal investigations.

When a user interacts with the CA they trigger a series of processes. Firstly, entities are extracted from their input text which can be later used in a database query. Next, the users input text is matched to a CA intention, which defines how the query will be structured, managed, and responded to. An intention is a collection of functional attributes and each of these has its own algorithmic processes and models. Additionally, the CA is interacting with a database, which has a defined structure and a degree of completeness and currency.

To date, much research into the area of XAI (Gunning 2019) has focussed upon ML systems for image classification, where there is a single model that is being explained with a single explanation. In these cases a typical approach is to identify the features in the image that are important in defining the classification, for example as a heat map over the image. Tools such as LIME (Ribeiro et al. 2016) allow users to inspect significant features visually, by highlighting them. These feature based approaches present what Gilpin et al. (2018) describe as a combination of interpretability and completeness, where interpretability is linked to explaining the internals of a system and completeness is to describe them as accurately as possible. The internals are captured by identifying and conveying the features within the input or output data itself. For example, if a user wishes to classify images of cats, a heat map layer that highlights the image pixels in the input images that are crucial to return the classification of 'cat', provides some explanation of the internals of the system.

Providing explanations becomes more complicated when applied to a real world system, such as our CA. As described previously, an interaction with a CA triggers various processes, including models for how to retrieve information. An explanation of the model used can be provided through the input or output data, to meet the needs described by Gilpin et al. (2018).

Table 1. Example Conversational Agent transcript (synthetic data).

Stage	Agent	Text entry
1	Human	Can you find people who are similar to Paul Peterson?
2	Machine	Yes, Sam Smythe is the most similar person to Paul Peterson. They are both employees of Dean Lane Services and have been offenders in criminal activities. Both have been involved in activities related to drugs. They both own vehicles

For example, in Table 1, the CA can respond and explain why 'Sam Smythe' has been chosen as similar to 'Paul Peterson'. It does so using features from data within the output, which are the entities and their relationships with one another. Previous research to develop an algorithmic transparency framework (Hepenstal et al. 2019) has identified the need to provide visibility of system goals and constraints, in additional to this explanation of the response data, so that a user can have a better understanding of when system outputs can be used. A further issue is that the explanation in Table 1 is related only to the information retrieval model used to return the response. If a user wanted to understand how to trigger an alternative intention, they could not grasp this from the explanation given because the intention classification involves an entirely different and separate system component. We would need to provide an additional feature based explanation for the ML classification of a user's intention, by highlighting the words that are important in the classification. With this simple example, we demonstrate that for real world systems explanations are not universally applicable, nor appropriate, across distinct system components and should therefore be tailored carefully.

To capture user needs for explanations of CA systems we conducted interviews with four intelligence analysts, each with more than 10 years of experience in an operational role. Our focus was upon using a CA for information retrieval tasks and interviews therefore covered a scripted investigation scenario, where an analyst is asking questions of a CA. Each interview lasted approximately an hour. We presented interviewees with a series of questions and corresponding CA responses with two explanation conditions. For one condition, responses described the data alone (Table 2) and in the other condition, responses described the data and the system process (Table 3). All data included in the text is fictional. For both conditions, we provided an accuracy measure for concept match confidence in the form of a numerical value between zero and one. We switched the order in which we presented conditions to analysts, so that two analysts saw condition 1 then 2, and two saw condition 2 then 1. We did this to help mitigate any ordering effects, for example, when we asked analysts to identify additional understanding needs or to compare the two conditions. We were not attempting to test the differences between the two conditions; rather we used the conditions as a starting point from which we could explore additional needs with the analysts.

Table 2. CA condition 1: Data only explanation

Q: Are any suspects connected to James White?
A: Yes, James White is connected to Frank Howes and Paul Keen

Table 3. CA condition 2: Data and system explanation

Q: Are any suspects connected to James White?
A: Yes, James White is connected to Frank Howes and Paul Keen James White was victim in an assault where Frank Howes and Paul Keen were suspects Concept match confidence = 0.92 I am looking for a connecting path between James White and suspects (Descriptor) I am looking for single shortest paths and therefore, will not consider more complex connections

We first reduced the interview transcript data to identify a total of 114 distinct statements made by the analysts, with counts for each analyst ranging from 24 statements to 34. To analyse the statements we used an approach called Emergent Themes Analysis (ETA), as described by Wong and Blandford (2004, 2002) where broad themes, which are similar ideas and concepts, are identified, indexed and collated.

Through rigorous coding of the data, we identified that analyst statements could be broadly associated with the core functional components of a CA that need explaining, for example entity extraction from the user's query or the system processes applied. This is an interesting finding, indicating that analysts have specific considerations for each function of a CA. We should not treat a user's understanding of the data in the same way as we provide an understanding of the intention classification, or the extracted entities, the response, or system processes. From broad CA component themes, we have drawn out the detailed understanding needed by an analyst in an explanation. These are the sub-themes and, in Table 4, we show that explanations can be tailored for each system component by placing greater emphasis on the relevant sub-themes. For example, system explanations require some understanding of the data, including clarification of source, currency and structure.

Table 4. CA component core understanding needs.

CA component theme	Sub-theme (common for multiple analysts)	Summary of sub-theme(s)
CA intention interaction	Clarification (3), Continuation (2)	Clear language to understand classification (i.e. no confusing response metric) and information to support continuation of investigation
Data	Clarification (3)	Clarification of data updates and source, and data structure to aid forming questions
Extracted entities	Clarification + Verification (3)	More information of entities extracted for clarification and verification
Response	Clarification (4), Justification (4), Exploration (2)	Justification of response with underlying data, clarification of language (not trying to be human) and terminology, ability to explore results in more detail
System processes	Continuation (4), Verification (4), Clarification (3), Exploration (2), Justification (2)	User wants system understanding to support continuation of investigation, to allow them to verify processes are correct and explore them in more or less detail and justify their use/approach and constraints

5 Conclusion

When the diversity of stakeholders and the complexity of AI-based systems with multiple components are taken into account, it is clear that a 'one-size-fits-all' approach to XAI is insufficient. Instead the particular needs of the target audience, be they stakeholders who influence the design of the system (developers) or those affected by the system (users), must be actively considered during the design process. Also, rather than focussing on single explanations to explain single models, specific system components may require different forms of explanation. These explanations should focus at both the global and local level, providing explanation of the individual outputs as well as the process by which they have been reached. Importantly, the system goals and constraints should be explained to both groups in order to develop an appropriate level of trust in the outputs. This will also aid recipients when deciding whether or not the system is appropriate to use within a given situation.

In summary, explanation needs in the real world are complex and context dependent. Ongoing work being conducted by the authors is focussed on understanding additional needs for explanations, for example when applied to different tasks, exploring the strengths and weaknesses of different approaches to explaining AI systems, methods for evaluating explanations, and the demonstration of a human-centred design approach to the development of XAI (specifically a conversational agent).

References

Blandford, A., Wong, W.: Situation awareness in emergency medical dispatch. Int. J. Hum Comput Stud. **61**, 421–452 (2004)

Doshi-Velez, F., Kim, B.: Towards a rigorous science of interpretable machine learning (2017)

Floridi, L., Cowls, J.: A unified framework of five principles for AI in society. Harvard Data Science Review, 1 (2019)

Gilpin, L.H., Bau, D., Yuan, B.Z., Bajwa, A., Specter, M., Kagal, L.: Explaining explanations: an overview of interpretability of machine learning. In: 2018 IEEE 5th International Conference on Data Science and Advanced Analytics (DSAA). IEEE, Turin (2018)

Gunning, D.: DARPA's explainable artificial intelligence (XAI) program. In: Proceedings of the 24th International Conference on Intelligent User Interfaces (IUI 2019) 2019 New York, NY, USA. Association for Computing Machinery, II

Hepenstal, S., Kodagoda, N., Zhang, L., Paudyal, P., Wong, W.: Algorithmic transparency of conversational agents. In: Trattner, C., Parra, D., Riche, N. (eds.) IUI 2019 Workshop on Intelligent User Interfaces for Algorithmic Transparency in Emerging Technologies, pp. 17–20. Los Angeles, CA, USA (2019)

Hoffman, R., Klein, G.: Explaining explanation, part 1: theoretical foundations. IEEE Intell. Syst. **32**, 68–73 (2017)

Hoffman, R., Miller, T., Mueller, S.T., Klein, G., Clancey, W.J.: Explaining explanation, part 4: a deep dive on deep nets. IEEE Intell. Syst. **33**, 87–95 (2018)

Lipton, Z.C.: The Mythos of model interpretability. In: 2016 ICML Workshop on Human Interpretability in Machine Learning (WHI 2016). New York, NY, USA (2016)

Mcneish, D., Maguire, M.: A participatory approach to helicopter user interface design. Ergonomics & Human Factors 2019. Stratford-upon-Avon (2019)

Mittelstadt, B., Russell, C., Wachter, S.: Explaining explanations in AI. In: FAT* 2019: Conference on Fairness, Accountability, and Transparency (FAT* 2019). Atlanta, GA, USA (2019)

Preece, A.: Asking "Why" in AI: explainability of intelligent systems-perspectives and challenges. Intell. Syst. Account. Finance Manage. **25**, 63–72 (2018)

Preece, A., Harborne, D., Braines, D., Tomsett, R., Chakraborty, S.: Stakeholders in explainable AI. In: AAAI FSS-18: Artificial Intelligence in Government and Public Sector. Arlington, Virginia, USA (2018)

Ras, G., van Gerven, M., Haselager, P.: Explanation Methods in Deep Learning: Users, Values, Concerns and Challenges. In: Escalante, H.J., et al. (eds.) Explainable and Interpretable Models in Computer Vision and Machine Learning. TSSCML, pp. 19–36. Springer, Cham (2018). https://doi.org/10.1007/978-3-319-98131-4_2

Ribeiro, M.T., Singh, S., Guestrin, C.: "Why should i trust you?": explaining the predictions of any classifier. In: KDD 2016 Proceedings of the 22nd ACM SIGKDD International Conference on Knowledge Discovery and Data Mining. New York (2016)

Samek, W., Wiegand, T., Muller, K.R.: Explainable Artificial Intelligence: Understanding, Visualizing and Interpreting Deep Learning Models (2017)

Tomsett, R., Braines, D., Harborne, D., Preece, A., Chakraborty, S.: Interpretable to whom? a role-based model for analyzing interpretable machine learning systems. In: 2018 ICML Workshop on Human Interpretability in Machine Learning (WHI 2018). Stockholm, Sweden (2018)

Weller, A.: Challenges for transparency. In: 2017 ICML Workshop on Human Interpretability in Machine Learning (WHI 2017). Sydney (2017)

Wong, W., Blandford, A.: Analysing ambulance dispatcher decision making: Trialing Emergent Themes Analysis. HF2002 human factors conference design for the whole person - integrating physical, cognitive and social aspects: a joint conference of the Ergonomics society of Australia (ESA) and the Computer human interaction special interest group, 2002 Canberra, Australia. Ergonomics Society of Australia

The Way We Think About Ourselves

Darshan Solanki, Hsia-Ming Hsu[✉], Olivia Zhao, Renyue Zhang, Weihao Bi,
and Raman Kannan

Tandon School of Engineering, New York, NY, USA
{das968,hmh371,olivia.zhao,rz1535,wb832,rk1750}@nyu.edu

Abstract. In the new normal of fake news, wide-scale disinformation and alternate facts, the need for fact-checks and bot detection is real and immediate. It is generally accepted that altering the psyche, our thinking, is the most potent form of controlling and shaping human behavior. Fake news, disinformation and alternate reality are all aimed at shaping our beliefs. In this experiment, we set out to understand if the stream of news articles is itself designed to influence society at large, either to think positively or negatively – in other words can dictate how society views itself – disconnected from reality. In this exercise we are seeking to identify if there is a systematic prevalence of positive/negative sentiment in a given stream of news articles, using standard NLP techniques.

Keywords: NLP · Sentiment analysis · Binary classifier · Disinformation · Fake-news

1 Cogito Ergo Sum

The often quoted, "I think, therefore I am" [1] is a profound reflection on human condition. To think is distinctly human and it is upon us to nurture and protect that faculty. Sometimes, our thinking springs forth from unknown source, as in the case of inspired works such as $E = mc^2$ [2] or Paradise Lost [3] and we don't need any protection from such sources. Then, there is, ephemeral source of information which are mostly rooted in some local context and short-lived, such as media in the myriad forms it is delivered to us. While it is up to the individual to choose wherefrom they source information, left unchecked, the potential for outlets, with undesirable objectives, to misrepresent reality, spread falsehood, mislead and shape societal thinking, is real and present. Arguably, Brexit in recent history and during World War II – an argument can be made that public opinion was shaped by a select few with access to media outlets.

1.1 Age of Disinformation and Fake-News

In the new normal of fake news, wide scale disinformation and alternate facts, the need for fact checks and bot detection is real and immediate. It is generally accepted that altering the psyche, our thinking, is the most potent form of controlling and shaping human behavior. Fake news, disinformation and alternate reality are all aimed at shaping our

© Springer Nature Switzerland AG 2020
D. D. Schmorrow and C. M. Fidopiastis (Eds.): HCII 2020, LNAI 12196, pp. 276–285, 2020.
https://doi.org/10.1007/978-3-030-50353-6_21

beliefs. In this experiment, we set out to understand if the stream of news articles is itself designed to influence society at large, either to think positively or negatively – in other words can dictate how society views itself – disconnected from reality. To answer this question, we have processed large number of content generated over a period of time. Each article was first prepared for NLP and then classified either as negative (depressing) or positive (uplifting) using several different classifiers. We then present a time series of the sentiment to understand if there has been a demonstrable shift in the sentiment of the article stream.

1.2 Technology Is a Double Edged Sword

In this new era of numerous technology advances such as, internet and social media tools, this problem is further exacerbated. So one can posit technology can amplify societal negative tendencies. However, other concomitant advances technologies such as machine learning, natural language processing, API driven access to data, allows us to devise solutions to counteract anti-social behaviors, and possibly mitigate this risk. The solution to a problem induced by technology, happens to be rooted in technology, as well.

This is urgent and in the rest of this paper we present our efforts to engineer a solution to classify individual articles using NLP and characterize streams of article and determine sentiment projected in s given stream of news articles.

2 Technical Overview

We performed a broad sentiment analysis of articles published by several digital outlets as a function of time and developed a time series of promoted sentiment for various media outlets. We do not know and we are not seeking to establish if the public opinion was indeed shaped during these periods. What we will establish is the sentiment article by article over time.

2.1 The Experiment

We processed articles from two outlets CNN and Guardian between Jan 2011 through Nov-2019 and retrieved 68158 articles from CNN and 38625 articles from Guardian. Our scope was to analyze one geographical region at a time. In this study we processed news from US region from both CNN and Guardian. Although we wanted to study news article from other outlets, these were the only two news corpus we could find.

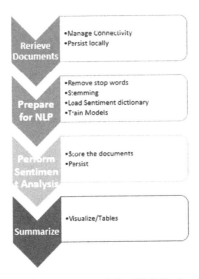

Each article once retrieved, was prepared for NLP Tasks, then we performed sentiment analysis, and each article was labeled as either positive or negative sentiment. This sentiment, the article, publisher and date of publication were persisted. This is a classic big data "pipeline" problem. Using this pipeline pattern, parallelizing is straightforward – each stage can be run in parallel using shared queue.

We now discuss the technical considerations and the architecture of our solution in detail for each pipeline stage. We implemented the peline in python.

2.2 Data Retrieval

Data Sources usually limit the rate at which clients retrieve data. Rate limits are imposed at the IP address level and/or api key level and sometimes on both IP address and api key. One must manage this tactfully so that we can complete a session in reasonable time, without being blocked by the content provider.

Source: *https://www.cnn.com/us/article/sitemap-2011-09.html*

```
class="sitemap-entry"><ul><li><span class="date">2011-09-30</span><span class="sitemap-link"><a href="https://www.cnn.com/2011/09/30/us/sport-florida-ramirez-charged/index.html"
Manny Ramirez charged with domestic violence</a></span></li><li><span class="date">2011-09-30</span><span class="sitemap-link"><a href="https://www.cnn.com/2011/09/30/us/radiohead-
street/index.html">Radiohead rumor swells Wall Street protest</a></span></li><li><span class="sitemap-link"><a href="https://www.cnn.com/2011
ceremony/index.html">Ceremony honors old, new Joint Chiefs chairmen</a></span></li><li><span class="date">2011-09-30</span><span class="sitemap-link"><a href="https://www.cnn.co
antibiotics/index.html">Authax antibiotics prescription plan needed, says report</a></span></li><li><span class="date">2011-09-30</span><span class="sitemap-link"><a
href="https://www.cnn.com/2011/09/30/us/same-sex-marriage-military/index.html">Military chaplains allowed to perform same-sex weddings</a></span></li><li><span class="date">2011
class="sitemap-link"><a href="https://www.cnn.com/2011/09/30/us/lettuce-recall/index.html">California farm recalls lettuce over contamination concerns</a></span></li><li><span c
class="date">2011-09-30</span><span class="sitemap-link"><a href="https://www.cnn.com/2011/09/29/us/cnnheroes-latino-top10/index.html">New hope for Chicago community &apos;plagued by violence&apos;</a></spa
<span class="date">2011-09-30</span><span class="sitemap-link"><a href="https://www.cnn.com/2011/09/29/us/latino-kids-poverty/index.html">Study: Largest U.S. group of poor kids is now
term</a></span></li><li><span class="date">2011-09-30</span><span class="sitemap-link"><a href="https://www.cnn.com/2011/09/30/us/scotus-preview-health-care/index.html">Health care, other hot issues pr
six days on leaves, water</a></span></li><li><span class="date">2011-09-29</span><span class="sitemap-link"><a href="https://www.cnn.com/2011/09/30/us/california-mountain-crash/index.html">Man stran
drops in overall percentage</a></span></li><li><span class="date">2011-09-29</span><span class="sitemap-link"><a href="https://www.cnn.com/2011/09/29/us/census/index.html">White U
convicted criminal immigrants arrested, ICE says</a></span></li><li><span class="date">2011-09-29</span><span class="sitemap-link"><a href="https://www.cnn.com/2011/09/28/us/immigration-sting/index.
fdny/index.html">Medal of Honor recipient declines city&apos;s offer to file late application for FDNY</a></span></li><li><span class="date">2011-09-29</span><span class="sitemap
href="https://www.cnn.com/2011/09/28/us/gitmo-trial/index.html">Guantanamo prepares for next military trial of terrorism suspect</a></span></li><li><span class="date">2011-09-29
```

source **view-source:https://www.cnn.com/us/article/sitemap-2011-09.html**

Using standard HTML parser that comes with Beautiful Soup package, we extract and store the URLs. Independently, content is retrieved from each URL using text libraries available in Python and it is stored in the file system as files, so we could leverage file processing capabilities and the meta data is persisted in the database with the following tuple structure:

Outlet: From which source we have gathered the data in our case CNN
Date: The published date of the article
Title: Title of that article:
Url: Actual URL of that article if someone wants to read it from the website
File_name

Guardian data is only marginally different allowing us to reuse much of the utilities we wrote for CNN.

Articles from Guardian are available from 2008 and CNN articles are available from 2011. There were lot more US articles from CNN as one would expect but by partitioning the tasks as described above, we distributed the load on 3 nodes and processed approximately 47 K articles in 80 min, at times processing more than 500 articles per minute.

```
df = pd.read_csv('Complete_Articles_Data.csv',names = ['outlet','date','title','url','text_file],sep='|')
df.tail 10
```

	outlet	date	title	url	text_file
69965	CNN	2019-4-30	USA Gymnastics director of sports medicine is	https://www.cnn.com/2019/04/30/us/usa-gymnast	AspBQGmdgSd7rWBwE5mGA1QjE8qg6bqN.txt
69966	CNN	2019-5-28	Ohio tornado survivor: It's heartbreaking	https://www.cnn.com/videos/us/2019/05/28/ohio-	pccnTgQw4svX1eQgnLn8GVJPRPBYUN.txt
69967	CNN	2019-4-29	The Illinois plant shooter threatened to kill	https://www.cnn.com/2019/04/29/us/aurora-illin	RrUrBJ7VgJCXs6kbb1OkJsHTSOVqAKSe.txt
69968	CNN	2019-5-1	New York is the first major city to allow free	https://www.cnn.com/2019/05/01/us/free-calls-f	pJYZgTH4TU148036CqecYnVn6UkRDZ.txt
69969	CNN	2019-5-1	A police officer responded to a noise complain	https://www.cnn.com/2019/05/01/us/police-offic	rvoZ6QVXXcVpZuqWKdkh5xPgIYx0dKg.txt
69970	CNN	2019-5-1	Maine becomes the first state to ban Styrofoam	https://www.cnn.com/2019/05/01/us/maine-ban-st	9q06ZWJXGZhZVWMDdiC6QwTat6y2IMrO.txt
69971	CNN	2015-8-25	John Kasich Fast Facts	https://www.cnn.com/2015/08/25/us/john-kasich-	OLYlUXggHawffloZU7wDHb7e8Dni7KpT.txt
69972	CNN	2019-5-1	Chicago sees slight drop in violent crime in A	https://www.cnn.com/2019/05/01/us/chicago-crm	Zi62b6QyKf2UsGTenOgiWd7k2M8Z30W.txt
69973	CNN	2019-5-1	Students stage walkout at Illinois high school	https://www.cnn.com/2019/05/01/us/blackface-il	wLWhyM4AhyQLJbC3F5opWW7750oCP.txt
69974	CNN	2019-5-1	2 Swarthmore fraternities will disband after d	https://www.cnn.com/2019/05/01/us/swarthmore-f	crbFZ8w60LmDKAhnaTyda/TubHpRWCCK.txt

We show here the scraping table.

2.3 Managing IP Address Using Proxies

As mentioned before, CNN limits us to 3500 calls per day and below we will elaborate the techniques we used to retrieve ~50 K articles in 80 min. We could not achieve the same level of throughput from Guardian perhaps because of internet latencies and possibly our public proxy ip addresses might have been shared.

2.4 Proxies

Exceeding the CNN limit, results in 24 h block period. To overcome this constraint we used Rotating IP service from US Proxy. After much trial and error, we chose to utilize paid proxy service so our proxies were not shared over the internet. We used a total of 50 IP addresses and 30 were valid.

The randomized proxy approach resulted in significant reduction in data gathering time.

2.5 Preprocessing

In this phase we removed duplicate articles, and performed the required NLP tasks, as follows:

1. removed extra spaces, special characters, single characters, new lines,
2. converted entire text to lowercase.
3. removed stop words using stop words library from nltk
4. performed lemmatization/stemming
5. removed participle, tense form of the words.

This preprocessing resulted in 50% reduction in the number of bytes to be processed.

2.6 Nlp

In this phase we classified each article using 5 different binary classifiers namely

1. Naive Bayes,
2. MultinomialNB,
3. BernoulliNB,
4. LogisticRegression, and
5. LinearSVC

and assigned a sentiment to the article using majority voting scheme.

2.7 Training Data for Sentiment Labeling

We use the known positive words and negative words to train our classifier models.
short_pos = open("trainning_files/positive.txt","r", encoding='iso-8859-1').read()
short_neg = open("trainning_files/negative.txt","r", encoding='iso-8859-1').read()

Words associated with positive sentiment

```
1979  wonder
1980  wonderful
1981  wonderfully
1982  wonderous
1983  wonderously
1984  wonders
1985  wondrous
```

Words associated with negative sentiment

```
4714  whore
4715  whores
4716  wicked
4717  wickedly
4718  wickedness
4719  wild
4720  wildly
```

We trained on 80% and tested on 20% of the data.

2.8 Verification and Testing

Let us consider the three sentences: "This article was rich, clear, willing, ingenuous, attractive, sensational, and hot"

"This is the best marvelous, imaginative, and realistic one I have seen"

"This article was utter junk. There were absolutely 0 points. I don't see what the point was at all. Horrible essay, sucks" with the corresponding result shown above.

```
['pos', 'pos', 'pos', 'pos', 'pos']
('pos', 1.0)
['pos', 'pos', 'pos', 'pos', 'pos']
('pos', 1.0)
['neg', 'neg', 'neg', 'neg', 'neg']
('neg', 1.0)
```

We applied this to entire document and for each document we tabulate the sentiment generated by the 5 classifiers as shown.

3 Result Analysis

We achieved an accuracy of 72% we got based on 80/20 split across the 5 classifiers as shown below

```
Original Naive Bayes model accuracy percent: 72.16494845360825
Most Informative Features
                    free = True          pos : neg    =      10.9 : 1.0
                   clear = True          pos : neg    =       8.6 : )
                  famous = True          pos : neg    =       5.5 : )
                    best = True          pos : neg    =       5.5 : )
                    safe = True          pos : neg    =       5.5 : )
                   sharp = True          pos : neg    =       5.5 : )
               effective = True          pos : neg    =       4.2 : )
              attractive = True          pos : neg    =       3.9 : )
                equivocal = True         pos : neg    =       3.9 : )
                  static = True          pos : neg    =       3.9 : )
                   noble = True          pos : neg    =       3.9 : )
             sensational = True          pos : neg    =       3.9 : )
                 envious = True          pos : neg    =       3.3 : )
                 willing = True          pos : neg    =       3.3 : )
                creative = True          pos : neg    =       2.4 : )
```

```
each classofoers average calculating time in sec:  [0.006393251199988299, 0.0019089575999787485,
0.0018411747999889485, 0.0021917007999800262, 0.0017657084000275063]
LogisticRegression_classifier accuracy percent: 72.38586156111928
LinearSVC_classifier accuracy percent: 72.82768777614137
MNB_classifier accuracy percent: 72.60677466863034
BernoulliNB classifier accuracy percent: 72.23858615611192
```

In the table below, the average executing time (around 16800 .txt files) and the accuracy achieved for each classifier is presented:

Classifier	NB	MultiNB	BinaryNB	Logistic	SVC
Accuracy percentage	72.16%	72.61%	72.24%	72.38%	72.83%
Executing time	0.00639	0.00191	0.00184	0.00219	0.00177

4 Visualization

Below we show number of positive/negative articles for CNN and Guardian for the entire period.

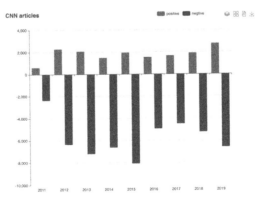

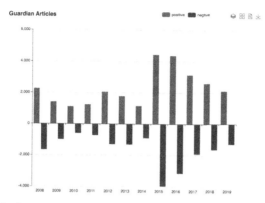

Sentiment analysis results of ten years.

4.1 Yearly Sentiment

In each year, the blue represents negative articles and the red represents positive.

We count the neg/pos articles in each month and we present the monthly neg/pos article count for all of 2019, using barcharts.

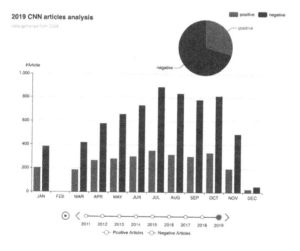

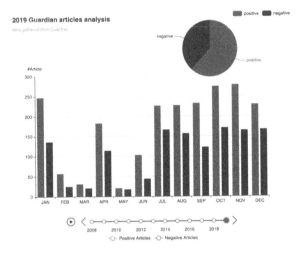

4.2 Sentiment Trend

In addition, we visualize the trend

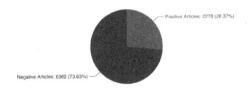

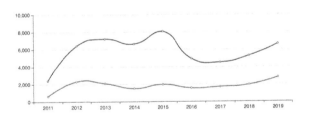

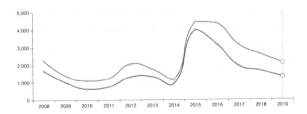

4.3 Production Deployment Urls

All work has been deployed using Microsoft Azure cloud and may be viewed here

1. https://newsarticlessentimentanalysis.azurewebsites.net/api/sentiment_engine?
   code=WXK9ko1U88HTrfB3oiOyFDsJDGpAa6JAuzLRjCNNkRoPInQNUqA
   SKw==&name=web.htm
2. http://newsarticlessentimentanalysis.azurewebsites.net/api/sentiment_engine?
   code=WXK9ko1U88HTrfB3oiOyFDsJDGpAa6JAuzLRjCNNkRoPInQNUqA
   SKw==&name=comparison.html
3. http://newsarticlessentimentanalysis.azurewebsites.net/api/sentiment_engine?
   code=WXK9ko1U88HTrfB3oiOyFDsJDGpAa6JAuzLRjCNNkRoPInQNUqA
   SKw==&name=fancy.html
4. http://newsarticlessentimentanalysis.azurewebsites.net/api/sentiment_engine?
   code=WXK9ko1U88HTrfB3oiOyFDsJDGpAa6JAuzLRjCNNkRoPInQNUqA
   SKw==&name=posvsneg.html

5 Conclusions

We find no discernible change in the positive/negative sentiment for CNN and Guardian as one would expect. We are actively seeking data to conduct additional experiments.

Acknowledgements. Generous support from IBM Power Systems Academic Initiatives (IBM PSAI) is acknowledged. Raman is grateful to the many determined volunteers and their fine contribution.

References

1. https://en.wikipedia.org/wiki/Cogito,_ergo_sum
2. https://en.wikipedia.org/wiki/Albert_Einstein
3. https://en.wikipedia.org/wiki/Paradise_Lost
4. Python newspaper documentation. https://buildmedia.readthedocs.org/media/pdf/newspaper/latest/newspaper.pdf
5. Text Summarization of an Article. https://medium.com/jatana/unsupervised-text-summarization-using-sentence-embeddings-adb15ce83db1
6. Insights about Nltk library. https://medium.com/datadriveninvestor/python-data-science-getting-started-tutorial-nltk-2d8842fedfdd
7. Usage of MultiCore Processing. https://medium.com/python-pandemonium/how-to-speed-up-your-python-web-scraper-by-using-multiprocessing-f2f4ef838686
8. How Lemmatization works. https://www.analyticsvidhya.com/blog/2018/02/natural-language-processing-for-beginners-using-textblob/
9. Data Source. https://www.cnn.com/us/article/sitemap-{yyyy}-{mm}.html, starting from 2011-07 till 2019-12-07
10. Data Source. https://www.theguardian.com/us-news/{yyyy}/{mm}/{dd}/all starting from 2008/jan/01 day till 2019/dec/31
11. Web Construction. https://getbootstrap.com/, https://www.echartsjs.com/en/index.html, and https://www.wix.com/
12. Data is uploaded on GitHub. https://github.com/Darshansol9/News_Articles_Sentiment_Analysis

Correction to: The Platonic-Freudian Model of Mind: Defining "Self" and "Other" as Psychoinformatic Primitives

Suraj Sood[✉]

Correction to:
Chapter "The Platonic-Freudian Model of Mind: Defining "Self" and "Other" as Psychoinformatic Primitives" in:
D. D. Schmorrow and C. M. Fidopiastis (Eds.): *Augmented Cognition*, LNAI 12196,
https://doi.org/10.1007/978-3-030-50353-6_6

In the originally published version of chapter 6, in section 4, an incorrect statement was made. This statement has been corrected as follows: "...System 1 consists of unconscious affect and System 2 consists of conscious cognition...".

The updated version of this chapter can be found at
https://doi.org/10.1007/978-3-030-50353-6_6

Author Index

Printed in the United States
By Bookmasters